Classical Algebra

Classical Algebra
Its Nature, Origins, and Uses

Roger Cooke
Williams Professor of Mathematics Emeritus
The University of Vermont
Department of Mathematics
Burlington, VT

WILEY-
INTERSCIENCE

A JOHN WILEY & SONS, INC., PUBLICATION

Published by John Wiley & Sons, Inc., Hoboken, New Jersey.
Published simultaneously in Canada.

For general information on our other products and services or for technical support, please contact our Customer Care Department within the United States at (800) 762-2974, outside the United States at (317) 572-3993 or fax (317) 572-4002.

Wiley also publishes its books in a variety of electronic formats. Some content that appears in print may not be available in electronic format. For information about Wiley products, visit our web site at www.wiley.com.

Library of Congress Cataloging-in-Publication Data:

Cooke, Roger, 1942–
Classical algebra : its nature, origins, and uses / Roger Cooke.
    p. cm.
  Includes bibliographical references and indexes.
  ISBN 978-0-470-25952-8 (pbk. : acid-free paper)
1. Algebra. 2. Algebra—History. 3. Algebraic logic. I. Title.
  QA155.C665 2008
  512—dc22
                            2007041610

Printed in the United States of America.

10 9 8 7 6 5 4 3

Contents

Preface   ix

Part 1.   Numbers and Equations   1

Lesson 1.   What Algebra Is   3
   1.   Numbers in disguise   3
     1.1.   "Classical" and modern algebra   5
   2.   Arithmetic and algebra   7
   3.   The "environment" of algebra: Number systems   8
   4.   Important concepts and principles in this lesson   11
   5.   Problems and questions   12
   6.   Further reading   15

Lesson 2.   Equations and Their Solutions   17
   1.   Polynomial equations, coefficients, and roots   17
     1.1.   Geometric interpretations   18
   2.   The classification of equations   19
     2.1.   Diophantine equations   20
   3.   Numerical and formulaic approaches to equations   20
     3.1.   The numerical approach   21
     3.2.   The formulaic approach   21
   4.   Important concepts and principles in this lesson   23
   5.   Problems and questions   23
   6.   Further reading   24

Lesson 3.   Where Algebra Comes From   25
   1.   An Egyptian problem   25
   2.   A Mesopotamian problem   26
   3.   A Chinese problem   26
   4.   An Arabic problem   27
   5.   A Japanese problem   28
   6.   Problems and questions   29
   7.   Further reading   30

Lesson 4.   Why Algebra Is Important   33
   1.   Example: An ideal pendulum   35
   2.   Problems and questions   38
   3.   Further reading   44

Lesson 5.   Numerical Solution of Equations   45
   1.   A simple but crude method   45
   2.   Ancient Chinese methods of calculating   46
       2.1.   A linear problem in three unknowns   47
   3.   Systems of linear equations   48
   4.   Polynomial equations   49
       4.1.   Noninteger solutions   50
   5.   The cubic equation   51
   6.   Problems and questions   52
   7.   Further reading   53

Part 2.   The Formulaic Approach to Equations   55

Lesson 6.   Combinatoric Solutions I: Quadratic Equations   57
   1.   Why not set up tables of solutions?   57
   2.   The quadratic formula   60
   3.   Problems and questions   61
   4.   Further reading   62

Lesson 7.   Combinatoric Solutions II: Cubic Equations   63
   1.   Reduction from four parameters to one   63
   2.   Graphical solutions of cubic equations   64
   3.   Efforts to find a cubic formula   65
       3.1.   Cube roots of complex numbers   67
   4.   Alternative forms of the cubic formula   68
   5.   The "irreducible case"   69
       5.1.   Imaginary numbers   70
   6.   Problems and questions   71
   7.   Further reading   72

Part 3.   Resolvents   73

Lesson 8.   From Combinatorics to Resolvents   75
   1.   Solution of the irreducible case using complex numbers   76
   2.   The quartic equation   77
   3.   Viète's solution of the irreducible case of the cubic   78
       3.1.   Comparison of the Viète and Cardano solutions   79
   4.   The Tschirnhaus solution of the cubic equation   80
   5.   Lagrange's reflections on the cubic equation   82
       5.1.   The cubic formula in terms of the roots   83
       5.2.   A test case: The quartic   84
   6.   Problems and questions   85
   7.   Further reading   88

Lesson 9.   The Search for Resolvents   91
   1.   Coefficients and roots   92
   2.   A unified approach to equations of all degrees   92

	2.1.	A resolvent for the cubic equation	93
3.		A resolvent for the general quartic equation	93
4.		The state of polynomial algebra in 1770	95
	4.1.	Seeking a resolvent for the quintic	97
5.		Permutations enter algebra	98
6.		Permutations of the variables in a function	98
	6.1.	Two-valued functions	100
7.		Problems and questions	101
8.		Further reading	105

Part 4.   Abstract Algebra   107

Lesson 10.   Existence and Constructibility of Roots   109
1.		Proof that the complex numbers are algebraically closed	109
2.		Solution by radicals: General considerations	112
	2.1.	The quadratic formula	112
	2.2.	The cubic formula	116
	2.3.	Algebraic functions and algebraic formulas	118
3.		Abel's proof	119
	3.1.	Taking the formula apart	120
	3.2.	The last step in the proof	121
	3.3.	The verdict on Abel's proof	121
4.		Problems and questions	122
5.		Further reading	122

Lesson 11.   The Breakthrough: Galois Theory   125
1.		An example of solving an equation by radicals	126
2.		Field automorphisms and permutations of roots	127
	2.1.	Subgroups and cosets	129
	2.2.	Normal subgroups and quotient groups	129
	2.3.	Further analysis of the cubic equation	130
	2.4.	Why the cubic formula must have the form it does	131
	2.5.	Why the roots of unity are important	132
	2.6.	The birth of Galois theory	133
3.		A sketch of Galois theory	135
4.		Solution by radicals	136
	4.1.	Abel's theorem	137
5.		Some simple examples for practice	138
6.		The story of polynomial algebra: a recap	146
7.		Problems and questions	147
8.		Further reading	149

Epilogue: Modern Algebra   151
1.		Groups	151
2.		Rings	154
	2.1.	Associative rings	154

2.2.	Lie rings	155
2.3.	Special classes of rings	156
3.	Division rings and fields	156
4.	Vector spaces and related structures	156
4.1.	Modules	157
4.2.	Algebras	158
5.	Conclusion	158

Appendix: Some Facts about Polynomials   161

Answers to the Problems and Questions   167

Subject Index   197

Name Index   205

Preface

My objective in writing this book was to help algebra students and their teachers to grasp the essence of classical polynomial algebra as a whole, to understand how it has developed and what it has developed into, to see the forest by looking at the trees. I have striven to answer questions such as the following: What is algebra about? How did it arise? What uses does it have? How did it develop? What problems and issues have arisen in its history, and how were those problems solved and those issues resolved? Since the chapters were originally very short, I preferred to call them "lessons," a name that I have retained as they grew longer in the rewriting.

I am mainly addresssing what seems to me to be a pedagogical disconnect between the subject taught as *algebra* in high school or as a remedial university-level course and the subject taught on the senior/graduate level in university courses called *modern algebra*. The typical high-school algebra course consists primarily of a set of rules for multiplying, dividing, and factoring polynomials, and unfortunately does not offer much explanation to the student about the ultimate usefulness of learning these techniques. At the other end of the spectrum, a course in modern algebra typically begins at a rather high level, with the abstract concept of a group, then progresses to rings, using polynomials as the primary example, and fields. At the end of this course the persevering student finally sees a connection between the two in the form of Galois theory. But there is a huge gulf between a quadratic equation and the concept of a Galois group. This gulf ought to make a person curious about the historical development that leads from the former to the latter.

The history of this development is rich in documents from ancient and medieval times showing what was achieved by Mesopotamian, Chinese, Hindu, ancient Egyptian, and Muslim scholars. Although I have never specialized in this area, I tried to describe it in general terms in my *History of Mathematics* (second edition, Wiley, 2005). But in writing history, one is constrained by the need to avoid anachronisms. It is an error to describe what a scholar did in terms of later, more successful efforts by other scholars, as if one were to say that Bach was trying very hard to write the kind of music Beethoven wrote. To write a pure history of algebra from ancient times to the year 1850 would require hundreds of pages.

Because my main interests are now in the history of twentieth-century physics, I had resolved to write no more general history of mathematics after finishing the second edition of my textbook. But when an invitation arrived from Amy Shell-Gellasch and Dick Jardine in January 2007 to write historical essays that teachers could use to supplement classroom presentations (the *Mathematical Capsules* project of the Mathematical Association of America), I could not resist the chance to say things in a slightly different way.

Attending only to my own agenda, I soon wrote much more than anybody could possibly use, and apparently in a style inconsistent with that of the others participating in the project. In the end, I submitted only two of the following lessons (alternative versions of Lessons 3 and 5 below) for the *Capsules* project. By that time, I was well into the writing, and decided to finish it. The result is the narrative that follows, a mixture of historical vignettes and elementary exposition of the main parts of polynomial algebra. As stated above, this book is aimed especially at teachers of algebra on all levels and also at students who wish to tie up the same loose ends that led me to write this book.

The "lessons" that follow do not constitute the *complete* story of algebra. The present work is mostly confined to the algebra of polynomials in one variable, and even in that narrow area, I have mentioned very few of the many authors and works that made this subject what it is today. Many mathematicians will probably be scandalized that I have written a book purporting to be a history of algebra without mentioning Cayley, Sylvester, Grassmann, and many others. Just how many contributors to the construction of the magnificent edifice of algebra have been slighted, their work callously and unfairly omitted from this account, can be judged by looking at more comprehensive histories written for mathematicians. For example, in the discussion of eighteenth-century developments, I have said very little about the work of Euler and mentioned only briefly certain parts of Lagrange's grand memoir on the solution of equations, ignoring the simultaneous and independent work of Vandermonde and Waring. For the interested reader, two good places to start filling in these gaps are the monographs by Luboš Nový, in the literature cited at the end of Lesson 9, and by Jean-Pierre Tignol, cited at the end of Lesson 11. The former, in particular, shows the role played in the genesis of modern algebra by the analysis of binary operations, whereas I have confined myself to the origins of group and field theory in the context of solving equations.

My excuse for omitting these people and topics is that I intend to discuss algebra in the sense it has for the average citizen, not as it is known to mathematicians. To do that, I have omitted almost everything not directly related to the algebraic solution of polynomial equations. The present book is close in spirit to the recent work of Peter Pesic, cited at the end of Lesson 10. It belongs to the genre that Grattan-Guinness calls *heritage*, focusing on "how things came to be the way they are" rather than "what happened in the past" (which is history). Those who are interested in knowing more of what was done in the past and what it looked like to contemporaries can read translations of the major works of algebra. English translations of the works of al-Khwarizmi, Umar al-Khayyam, and Girolamo Cardano, for example, do exist.

Compared to present-day mathematicians, these early algebraists were groping in the dark. The dawn came very slowly, and it was many centuries before polynomial equations were seen in the clear light of day. Once the dawn has come, it would be foolish to close the curtains and go back to groping in the dark.

Outline of the book. The first four lessons investigate the nature and importance of algebra as it is now taught to high-school students and some first-year university students. Lesson 5 presents the highlights of numerical solution of equations. Lessons 6 through 11 stay somewhat closer to the historical development of the subject that I call the formulaic solution of equations. A rough division of this development into three periods is furnished by the different conceptual approaches that were tried and pushed to their limits, then supplemented by new techniques. The first phase, which I refer to as the *combinatorial* period, involves the use of substitutions to reduce an equation to a form in which algebraic identities allow it to be solved by extracting roots; this period is discussed in Lessons 6 and 7 and ends with the Cardano solution of the cubic equation. The next phase involves the Tschirnhaus solution of the cubic and the solution of the quartic equation, both of which bring to light a kind of bootstrapping process, whereby substitutions are sought that allow the degree of the equation to be reduced. Particularly important is the concept of a *resolvent*, the dominant theme in the second phase, which I naturally call the *resolvent* period. It is discussed in Lessons 8 and 9. Finally, the search for a resolvent of the general quintic led to the creation of abstract algebra, beginning with the study of the permutations of the roots and their effect on hypothetical resolvents and finally resulting in proofs that no algebraic solution of the general quintic exists (Lesson 10) and a general method of analyzing equations (Galois theory, discussed in Lesson 11) to see whether their solutions can be expressed as algebraic formulas. This phase of the subject continues today, a full two centuries later. I call it the period of *modern algebra*.

As an Epilogue, I discuss very briefly some of the central concepts of modern algebra as it has been taught for the past century.

Prerequisites. Although I had originally called these essays "easy lessons," they are not all equally easy, and all of them have gotten harder as one draft has succeeded another. Although I explain some of the undergraduate curriculum, especially linear algebra, on a need-to-know basis, the exposition is not systematic, and some core topics are used without proof. I regard linear algebra as the cleanest subject in the undergraduate mathematics curriculum and hope that the reader who has not yet had this course will be patient and take such a course as soon as possible. Three other topics that I refer to (rational roots of equations, the Euclidean algorithm, and Descartes' rule of signs) are discussed in the Appendix.

Beyond the linear algebra just mentioned, the main requirement for reading the first nine lessons is the ability to add and multiply simple polynomials, which is one of the early skills taught in algebra. It will also help if the reader has at least heard of imaginary and complex numbers. I am assuming that some of my readers will have had only one year of algebra, but that others may have gone on to study calculus and even modern algebra on the university level. Consequently, at a few points, I invoke some more advanced topics such as trigonometry, differential equations, elliptic functions, and vectors and vector spaces without explaining what these things are. These passages can be omitted by the reader who is not yet familiar with them. I believe the parts of the book that are accessible to the average university undergraduate or high-school student will still be worth the reader's time.

Although the main ideas of this book can be followed without knowing much advanced algebra, I am alerting the reader here that some rather formidable-looking mathematics pops up occasionally, even in the early chapters, in the form of field extensions, quaternions, and so forth. I implore the unsophisticated reader to skim over these rough spots, which are included in many cases only as examples. I believe the essence of the story of algebra can be understood without these details, and I hope that the reader will return and read them again, after getting some help from people who have studied these topics in formal courses.

The last two lessons, however, do make heavy demands on the reader's patience and sophistication. Here my opportunistic use of snatches of group theory with only minimal explanation would be outrageous in a textbook. My excuse for introducing this topic is twofold. First, some of my readers, I hope, will already know what these things are, and will be able to appreciate my condensed explanation of Galois theory. Second, those readers who have not studied group theory may still be able to understand the essence of what I am saying, and may be inspired to undertake a systematic study of this rewarding area of mathematics. Minimal explanations of all these concepts are offered in the Epilogue and Appendix.

I am grateful to Amy Shell-Gellasch and Dick Jardine for getting me started on this book, and I would like to express special thanks to Garry J. Tee, who at the last moment sent me a list of corrections and suggestions that have greatly improved the result. I am, of course, the only one responsible for the defects that remain.

Roger Cooke   *December 9, 2007*

Part 1

Numbers and Equations

The first five lessons consist of general information and reflections on numbers and equations and the meaning of algebra. Lessons 1 and 2 discuss the relation between arithmetic and algebra. Lessons 3 and 4 inquire into the value of algebra for science and human culture as a whole. Lesson 5 is devoted to the numerical approach to solving equations, as opposed to the formulaic approach that will be our main concern in the rest of the book.

LESSON 1

What Algebra Is

In these lessons, we are going to explore key moments in the development of algebra in different places over the past 3500 years. As we shall see, different people have written about algebra in different ways, depending on the kinds of problems they were solving and the ways in which they manipulated numbers. In order to get a perspective that will enable us to appreciate what all these writings have in common, we devote this first lesson and the one following to some very general considerations. In the present lesson, we explore the nature of algebra itself and the different number systems in which its problems are stated and solved.

1. Numbers in disguise

As human societies grow larger, their administrative complexity grows disproportionately. While a single leader can make all the decisions on where to hunt, where to encamp, how to watch out for enemies, and so on for a small clan in which everyone knows everyone else, large societies, in which people must often deal with strangers, require formal laws to govern behavior. As economies become more complex, it is necessary to regulate commerce, weights, and measures and to plan strategically for defense or conquest. Over time, a group of specialized bureaucrats arises, charged with administering these vital activities.

These bureaucrats universally rely on two forms of mathematics: arithmetic and geometry. To collect taxes on land, to regulate trade and agriculture, to design and construct large public works, it is vital to know the elements of these two subjects. Records show that the people of Egypt and Mesopotamia possessed this knowledge at least 4000 years ago. Undoubtedly, such knowledge was also current in China and India about the same time. However, there is evidence that the Chinese used mechanical methods of calculating, in the form of counting rods, rather than graphical methods, and thus the details of their mathematics have vanished. Whether for that reason or because the first Emperor Ch'in Shih Huang Ti ordered the burning of all books when he unified China in 221 BCE, only a few Chinese texts known to be more than 2000 years old have been preserved.

Although the term *bureaucrat* has an unfortunate connotation that suggests a soulless automaton, mindlessly enforcing rules, the bureaucrats of these early societies were, like all human beings, possessed of an imagination, and they were the first people who were given economic support that enabled them to indulge their imagination. They must have been encouraged to plan strategically; not only to see that the current year's harvest is properly stored, distributed, and taxed but also to consider the possibilities of external aggression, future bad weather, and the like. If they were asked to design monuments, bridges, roads, and tunnels, such tasks would exercise their imaginations.

Perhaps in the intervals of their administrative work they found time to play games with the mathematical knowledge they possessed, posing problems for one another. This last activity may well explain why the earliest texts contain so many examples of problems for which a practical application is difficult to imagine. Or perhaps the explanation is our own lack of imagination about the kinds of practical problems they actually faced. Whichever is the case, we find arithmetic and geometry combining in many of these early texts to produce what we might call mathematical riddles, or perhaps *numbers in disguise*, with a challenge to unmask the numbers and make them reveal themselves, as in the following fictional anecdote.

Example 1.1. The dynasty of Uresh-tun was the wonder of its neighbors because of the prodigiously tall tree that grew just outside the walls of the king's castle. The kings of this dynasty held court under its branches in pleasant weather. No one knew what kind of tree it was; there was none like it for hundreds of miles around. Then, during the reign of the seventh king of the dynasty of Uresh-tun, this marvelous tree was blown over by a storm and fell with a great crash. The king commanded that it be cut into planks for his own use, and this was done. The largest of these planks was perfectly straight and of even thickness throughout and measured 44 meters in length and 75 centimeters in width. What suitable use could the king make of such a treasure? It was too long to fit inside any of his buildings, and he did not wish to leave it outside to rot in the damp weather.

After much thought, he decided on a use for it: It would furnish the frame for a set of portraits of himself and his six illustrious predecessors of the dynasty. He summoned his artisans and ordered them to cut notches at the ends and at three other points in such a way that the four pieces would provide a single frame for seven identical square tiles on which the portraits would be painted.

The artisans recognized that they must cut out three isosceles right triangles at three points on the plank and two others half as large, one at each end. Where should the three interior cuts be made? Obviously, one of them should be exactly in the middle. But where should the other two go? They could see that removing the triangles at the two ends would decrease the perimeter of the inside of the frame by 1.5 meters, and each of the other three cuts would remove another 1.5 meters, so that the rectangular inside of the frame would have a perimeter of 38 meters. The problem was to make that inner rectangle seven times as wide as it was high.

The folk wisdom of Uresh-tun said, "Measure twice before cutting once," and they knew that the king would not forgive any bungling on their part.

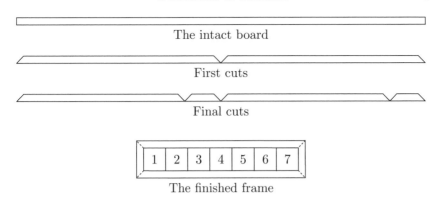

FIGURE 1. Cutting a board to make a picture frame.

They dared not experiment on such a precious piece of wood, and there was no other piece of such length on which they could make practice cuts. They had to get it right the first time. Symmetry showed them that the left and right halves of the plank would have to be cut identically. The problem that remained was to divide a length of 19 m into two parts so that one of the parts was seven times the other.

That is where we leave the artisans. You may enjoy thinking of both experimental and computational ways by which they might solve this problem. Probably you will agree that the computational way is somehow "neater" and more satisfying than trial and error, and much faster, once you see how to do the problem. To visualize it, look at Fig. 1.

Having seen first-hand in histories of mathematics how easily urban legends and folk tales begin, I do not wish to be the source of any new ones. Hence I emphasize again that this example is pure fiction. As far as I know, there has never been any place called Uresh-tun anywhere, much less one that generated the problem just described. However, the pure mathematics problem that corresponds to it was stated by an Egyptian scribe nearly 4000 years ago: *A quantity and its seventh part together equal* 19. *What is the quantity?*

If you wish to see how the Egyptian scribe solved this problem, look ahead to Lesson 3. However, try to solve it yourself, by both practical and mathematical means. There are several ways to proceed. (See Problem 1.11.)

1.1. "Classical" and modern algebra. The carpentry problem just posed leads to a single linear equation in one unknown. As such, it can be solved by pure arithmetic, and so marks the borderline between arithmetic and algebra. You don't *have* to introduce an equation to solve this problem, although you can if you wish.

What this kind of problem reveals is that numbers do not have to be named explicitly in order to be determined. They are sometimes determined by properties that they have. This way of thinking can apply to any objects, not just numbers. In geometry, lines are often determined by certain properties, such as being tangent to a circle at a given point. The technique of thinking in terms of descriptions, which is the essence of the early algebra we will be describing, was mentioned by the fourth-century geometer Pappus of Alexandria (ca. 290–ca. 350) in Book 7 of his *Collection*. After explaining that analysis proceeds from the object being sought to something that was agreed on (known to be true), he said, "For in analysis we set down the object being sought as something that has been constructed, and then examine what follows from this; then we repeat with that consequence, until by such considerations we arrive at something either already known or some first principle." He was thinking of geometric objects, but his *analysis* reflects the same kind of thinking used in algebra, where we write down a symbol for the unknown number as if it were already at hand, and then consider the conditions that it must satisfy. In our board-cutting example, the unknown number is characterized as being seven times the difference between 19 and the number itself.

The technique described by Pappus lies at the heart of even the more advanced and subtle thinking involved in the general solution of polynomial equations. Although no general method for finding the roots of a fifth-degree equation was known in the early nineteenth century, nevertheless mathematicians could write down five symbols to represent those roots and reason about the properties they must have. The result was eventually a proof that no finite algebraic formula expressing them exists.

Thus, numbers may appear in disguise, and this way of thinking about them forms the subject that we are going to call *classical algebra*. By that term, we mean the algebra that was practiced in many parts of the world for about 4000 years, from the earliest times to the midnineteenth century. This algebra was confined to the study of polynomial equations, an example of which is the quartic (fourth-degree) equation

$$x^4 - 10x^3 + 3x^2 + 2x - 7 = 0\,.$$

By the year 1850 the major questions in classical algebra had received answers, and that is the portion of the story of algebra that will be told in this book.

When difficult mathematical problems that have been open for a long time are finally solved, the techniques that were used to solve them generate their own interesting questions and become the foundation of a new subject. In this case that subject is known as *modern algebra*, and it studies general operations on general sets. The most important structures of this type are called *groups*, *rings*, *fields*, *vector spaces*, *modules*, and *algebras*, which are vector spaces whose elements can be multiplied. The most abstract form of algebra, known as *universal algebra*, studies arbitrary unspecified classes of operations satisfying certain laws, all of which are generalizations of the familiar properties of numbers.

For the sake of perspective, we describe parts of modern algebra briefly in the Epilogue that follows Lesson 11. We will have to invoke some of the concepts of modern algebra toward the end of the story of classical algebra, but for the first nine lessons, we can avoid most of them. The only concept we will make constant use of is that of a *field*, described below. Having now defined our subject matter, we shall henceforth drop the adjective *classical*, with the understanding that when we refer to algebra, we mean the topic of polynomial equations unless we state otherwise.

2. Arithmetic and algebra

Most people would probably describe the difference between algebra and arithmetic by saying that in algebra we use letters in addition to numbers. That is a fair way of telling the two subjects apart, but it does not reveal the most important distinction between them. Letters are a convenient notation for recording the processes that we use in algebra, but algebra was being done for some 3000 years before this notation became widespread in the seventeenth century. With a few exceptions such as the Jains in India, who used symbols to represent unknown numbers, the earliest authors wrote their algebra problems in ordinary prose. When you see problems written in prose, it can be more difficult to distinguish between algebra and arithmetic. In both cases, you are given some numbers and asked to find others. What then is the real difference? Let us look at an example to make it clear.

An arithmetic problem: $3 \times 7 + 36 = ?$

An algebra problem: Solve the equation $3x + 36 = 57$.

Let us see what these two problems look like when stated in prose. In the first problem, we are given three numbers (data), namely 3, 7, and 36. We are also given certain processes to perform on these numbers, namely to multiply the first two, then add the third number to the product. We get the answer (57) by following the known rules of arithmetic. Arithmetic amounts to the application of addition, subtraction, multiplication, and division to numbers that are explicitly named.

In the second problem, we are presented with an unknown number. We are told that when it is multiplied by 3 and 36 is added to the product, the result is 57. We must then find the number. As you can see, the biggest difference here is that we are not told what processes we must use in order to find the unknown number. Instead, we are told that some arithmetic was performed on a number, and we are told the result.

Schematically, we are looking at the same underlying process in both cases:

$$(\text{data}), (\text{arithmetic operations}) \longrightarrow (\text{result}).$$

In arithmetic we get the data and the operations given to us and must find the result. In algebra, we get the operations and the result and must find the original data.

This difference can be illustrated by analogies from everyday life. The problems that come to us in algebra are a challenge to find concealed numbers. The equations in which they occur are like locked boxes containing valuables. A technique for solving them is like a key to open the box. To take a different analogy, an equation is like a chunk of ore from a mine. The minerals it contains are all jumbled together. It takes a chemist to determine what those minerals are and a metallurgist to separate them so that they can be used. This analogy is better than the first, since chemists and metallurgists study the ways in which minerals combine in order to understand how to separate them again. In the same way, algebraists study the ways in which numbers combine in order to find techniques for separating them, and the study of chemistry or algebra is a perfectly respectable occupation in itself, independently of any minerals or numbers that one may eventually extract from a piece of "ore."

3. The "environment" of algebra: Number systems

The data in an equation and its solutions are numbers. But what kind of numbers are they to be? To solve linear problems like the equation $3x + 36 = 57$ given above, we need only the operations of arithmetic. However, in order to perform these operations, we must have a sufficiently general set of numbers to work with. The positive integers work fine for addition and multiplication. But to make subtraction possible, we need to adjoin zero and the negative integers. Then, to make division (except by zero) possible, we also need to allow all proper and improper fractions. For that reason, the smallest set of numbers that we could possibly consider reasonable would be the *rational numbers* (all fractions, positive and negative, proper and improper). For later reference, we note that a number system in which the four operations of arithmetic are possible, with the exception of division by zero, is called a *field*. For brevity, the four operations of arithmetic are referred to as the *rational* operations. Rational operations can always be performed *within* a field, without adjoining any new elements. In contrast, root extractions are not always possible, and fields must sometimes be enlarged to accommodate them. In fact, the process of enlarging fields by adjoining roots lies at the very heart of the problem of solving equations. Expressions formed using a finite number of rational operations and root extractions are called *algebraic formulas*.

In the present lesson, we shall encounter four fields: the rational numbers, the real numbers, the algebraic numbers, and the complex numbers, all defined below. But there are many others, including some finite fields of considerable interest in algebra, which we shall explore in the problem set below. Let us start with the smallest of these four fields, the rational numbers, which we shall always denote $\mathbb{Q}$. These numbers are not sufficient for solving all equations. To solve an equation like $x^7 = 10$, we must be able to extract roots as well, and this operation forces us to consider a larger class of numbers, in which root extractions are possible. We shall refer to these five operations from now on as the *algebraic operations* on numbers.

Allowing root extractions forces us to include certain *irrational* numbers in our set of possible solutions since, for example, $\sqrt{2}$ is not a rational number. We might even want to solve $x^2 = -1$, and so we shall also include *imaginary* and *complex* numbers. A complication arises when we allow root extractions, since every complex number except 0 has exactly two square roots, three cube roots, and so on. For example, the fourth roots of -4 are $1 + i$, $1 - i$, $-1 + i$, and $-1 - i$, where $i = \sqrt{-1}$. Thus, when we extract a root, we must either decide which of the possible roots we want, or else live with a symbol representing more than one number. Any complex number that is not a rational number is called an *irrational* number, but the term is most often applied to real numbers that are not rational.

To avoid having to invent new numbers all the time, we need a field that contains the rational numbers and is such that every equation with coefficients in the field will also have a solution in the field. Such a field is called *algebraically closed*. The smallest algebraically closed field is called the set of *algebraic numbers*. This field includes all roots of integers, even roots of negative integers, so that some complex numbers, such as the fourth roots of -4 listed above, are algebraic. To be precise, an algebraic number is any number (real or complex) that satisfies an equation whose coefficients are rational numbers. If you have such an equation, you can multiply it by a common multiple of the denominators of the coefficients and get an equation having the same roots, but with integer coefficients. For example, the equation $\frac{3}{2}x^2 - \frac{5}{7} = 0$ is equivalent to $21x^2 - 10 = 0$. Thus, the phrase *rational numbers* in the definition of an algebraic number could have been replaced by the word *integers*.

Algebraic numbers include all numbers that can be formed starting from rational numbers using a finite number of our five classes of operations, for example, $\sqrt{2} + \sqrt[3]{2}$. Since there are two square roots of 2 and three cube roots of 2, this expression might represent any of six numbers. Because these six numbers are algebraic, they must be roots of an equation with integer coefficients. You can verify that they are in fact the six roots of the equation

$$x^6 - 6x^4 - 4x^3 + 12x^2 - 24x - 4 = 0.$$

Remark 1.1. Throughout these lessons, we may use the word *root* to refer to a value of x that makes a polynomial $p(x)$ equal to zero ("root of the polynomial") or to a value of x that makes a polynomial equation $p(x) = 0$ true ("root of the equation") or to a complex number, some power of which equals a given complex number z ("root of the complex number z", that is, a root of a polynomial $x^n - z$, which is the same as a root of the equation $x^n = z$).

Obviously, algebraic numbers can easily acquire a very messy appearance, for example,

$$\frac{\sqrt[3]{37} + \sqrt{21}}{4 + \sqrt[5]{19}} - \sqrt{\frac{-9}{7}}.$$

Thinking of a messy expression like this is a good way to picture a "typical" algebraic number. Because of the ambiguity of taking roots, this expression actually represents 60 different complex numbers! In practice, we would probably choose the simplest of the possible values, which is approximately $1.36415 - 1.13389i$.

Remark 1.2. A few words of caution are needed here. Although we have just told the reader to think of an algebraic number as a finite expression involving rational numbers, arithmetic operations, and root extractions, numbers of this form are very far from being typical algebraic numbers in an abstract sense. Not every algebraic number can be written as the result of applying a finite number of arithmetic operations and root extractions to rational numbers. In other words, not every algebraic number can be written as a *formula* involving only rational numbers, arithmetical operations, and root extractions. For example, the five roots of the equation $x^5 - 10x + 2 = 0$ cannot be written this way. This impossibility can be proved using Galois theory, an invention of Évariste Galois ("GAL-wa," 1811–1832) and the earliest achievement of modern algebra. In Lesson 11, we shall sketch a proof of this impossibility.

As far as algebra itself is concerned, algebraic numbers would be sufficient for all needs. However, many algebra problems arise from applications in geometry, and these are quite likely to involve the number π, which is *not* an algebraic number. Nonalgebraic complex numbers are called *transcendental* numbers, since they "transcend" algebra. Every transcendental number is irrational, but most of the common irrational numbers are algebraic rather than transcendental. Because of the applications in geometry, it is simplest just to take the whole set of real and complex numbers as the set in which we seek solutions of our equations. For our purposes, a real number is a finite or infinite decimal expansion, and a complex number is a number of the form $a + bi$, where a and b are both real numbers and $i^2 = -1$. It happens to be true that every equation with coefficients in this set will also have a solution in the complex numbers. In other words, like the algebraic numbers, the complex numbers form an algebraically closed field, one that is larger than the algebraic numbers.

Remark 1.3. Although the complex numbers are used in algebra, and indeed essential in the subject known as algebraic geometry, they are much more geometric than algebraic in nature. The difference between the algebraic and the analytic construction of numbers is well illustrated by the number $\sqrt{2}$. In real and complex analysis, this irrational number can be located as the point where the circle through the point $(1, 1)$ with center at $(0,0)$ intersects the positive real axis. The proof that there actually is a point of intersection involves the order axioms of geometry. For algebraists, a number whose square is 2 is constructed as part of an extension of the field $\mathbb{Q}$ of rational numbers to a larger field denoted $\mathbb{Q}(\theta)$. This larger field consists of formal expressions $r + s\theta$, where r and s are rational numbers. Addition of such pairs follows the usual algebraic rules, and multiplication is defined by $(r + s\theta) \times (t + u\theta) = (rt + 2su) + (ru + st)\theta$. For example, $(2 - 3\theta) \times (1 + 5\theta) = -28 + 7\theta$. The field $\mathbb{Q}$ of rational numbers is identi-fied with a subfield of $\mathbb{Q}(\theta)$ via the "injection" mapping $r \mapsto r + 0\theta$. This injection preserves addition and multiplication, and so makes it reasonable to identify $\mathbb{Q}$ as a part of $\mathbb{Q}(\theta)$. Then the number $\theta = 0 + 1\theta$ satisfies $\theta^2 = 2 + 0\theta = 2$, so that θ amounts to a square root of 2.

This algebraic process for constructing $\sqrt{2}$ is finite, requiring no geome-try or approximating processes. Contrast this finiteness with the construc-tion of this number used by analysts. As a real number, $\sqrt{2}$ requires infinite precision to define, either as the infinitely small point on the intersection of the line and circle mentioned above, or as the infinite decimal expansion $\sqrt{2} = 1.41421\ldots$, which never repeats and never ends.

The distinction between "finite" algebra and "infinite" or "infinitesimal" (infinitely small) analysis made here is not absolute. As already pointed out, not every algebraic number can be written as a formula involving only a finite number of algebraic operations and rational numbers. Even algebra resorts, at some point, to potentially infinite processes.

Remark 1.4. It can be difficult to determine whether a complex number is algebraic. Except for certain artificially constructed examples, the decimal expansion of an irrational number seldom helps to determine whether the number is algebraic or transcendental. Not until the nineteenth century were mathematicians able to prove, for example, that the fundamental constants $\pi = 3.14159...$ and $e = 2.71828...$ are transcendental.

4. Important concepts and principles in this lesson

Before proceeding to the next section, be sure you have a clear picture of each of the following concepts: equation, unknown, coefficient, integer, rational number, rational operation, algebraic number, algebraic formula, transcendental number, real number, and complex number.

As you continue reading, keep in mind the analogies we have introduced here, comparing algebra to the analysis of an ore or the unlocking of a sealed box. Here is another that may help: Doing arithmetic is like cooking; you follow a recipe using specified ingredients processed using available machin-ery. Doing algebra is like being a food taster; you try to find out what the original ingredients were by looking at the final result.

5. Problems and questions

Problem 1.1. It is possible to make a field out of as few as two elements, which must necessarily be 0 and 1 and must have the following tables for addition and multiplication:

+	0	1
0	0	1
1	1	0

×	0	1
0	0	0
1	0	1

Show that subtraction is the same as addition in this field. That is, the equation $x + a = b$, which should have the solution $x = -a + b$, actually has the solution $x = a + b$. There are only four possible sets of values (a, b). Verify this for all four possible choices. In fact, these tables are merely the rules for manipulating even (0) and odd (1) integers, that is, even times odd equals even, and so on.

Problem 1.2. Solve the quadratic equations $x^2 + 1 = 0$ and $x^2 + x = 0$ in the two-element field just exhibited, and show that the equation $x^2 + x + 1 = 0$ has no solution at all.

Problem 1.3. There is also a field having exactly three elements, which we shall label 1, -1, and 0. Its addition and multiplication tables are as follows:

+	0	1	-1
0	0	1	-1
1	1	-1	0
-1	-1	0	1

×	0	1	-1
0	0	0	0
1	0	1	-1
-1	0	-1	1

In this field we have the strange-looking rules $1 + 1 = -1$ and $(-1) + (-1) = 1$. This would look less strange if we adopted a more familiar notation and wrote 2 instead of -1. If we did that, the strangeness would be expressed in the rules $1 + 2 = 0$ and $2 + 2 = 1$. Show that these rules are merely the common rules for manipulating the remainders when integers are divided by 3. For example, if each of two numbers leaves a remainder of 2 (that is, -1) when divided by 3, then their sum and product each leave a remainder of 1. (In other words, $2 + 2 = 4$ and $2 \times 2 = 4$, as usual; only $4 = 1$ in this case, since the remainder when 4 is divided by 3 is 1.)

Problem 1.4. What does subtracting 1 or -1 mean in the three-element field? What does dividing by these elements mean?

Problem 1.5. Find all the quadratic equations that can be solved in the three-element field, and write down one that cannot be solved.

Problem 1.6. There is a field having exactly four elements, but its arithmetic is not analogous to the arithmetic of the fields with two and three elements. We reserve discussion of this field for Lesson 6 below. Instead, at this point, we introduce the five-element field, whose elements are -2, -1, 0, 1, and 2. (Or, if we prefer, 0, 1, 2, 3, and 4. In any case, it is just the arithmetic of remainders after division by 5.) Its addition and multiplication tables are as follows:

+	0	1	−1	2	−2
0	0	1	−1	2	−2
1	1	2	0	−2	−1
−1	−1	0	−2	1	2
2	2	−2	1	−1	0
−2	−2	−1	2	0	1

×	0	1	−1	2	−2
0	0	0	0	0	0
1	0	1	−1	2	−2
−1	0	−1	1	−2	2
2	0	2	−2	−1	1
−2	0	−2	2	1	−1

What does the fraction $1/2$ mean in this field? (*Hint:* It should be a solution of the equation $2x = 1$.)

Problem 1.7. The complex number $x + iy$ is naturally identified with the point (x, y) in the plane. Attempts by William Rowan Hamilton (1805–1865) to regard "vectors" (x, y, z) in three-dimensional space as part of a field, on which rational operations could be performed, were unsuccessful, until he embedded them in a larger four-dimensional space of vectors (t, x, y, z). Hamilton named this four-dimensional system *quaternions*. It will be described in the next problem. After that, Josiah Willard Gibbs (1839–1903) was able to distill an algebraic system for three-dimensional vectors by multiplying them as quaternions and projecting them back onto the last three coordinates (the cross product) or the first coordinate (the dot product). For more on vectors in general, see the Epilogue. If $\alpha = (a_1, a_2, a_3)$ and $\beta = (b_1, b_2, b_3)$, their cross product is $\alpha \times \beta = (a_2 b_3 - a_3 b_2, a_3 b_1 - a_1 b_3, a_1 b_2 - a_2 b_1)$. The vectors α and β also have an "inner" or "dot" product that is a number rather than a vector: $\alpha \cdot \beta = a_1 b_1 + a_2 b_2 + a_3 b_3$.

Verify the following simple facts:

1. $\alpha \cdot \alpha = a_1^2 + a_2^2 + a_3^2$. This number is obviously positive unless $\alpha = (0, 0, 0)$. Its square root is called the *norm* or *length* or *absolute value* of α and denoted $|\alpha| = \sqrt{\alpha \cdot \alpha}$.
2. $(\alpha \cdot \beta)^2 \le (\alpha \cdot \alpha)(\beta \cdot \beta)$. This inequality is called the *Schwarz inequality* after Hermann Amandus Schwarz (1843–1921). This is obvious if $\alpha = (0, 0, 0)$. In all other cases, consider the vector $\gamma = (\alpha \cdot \beta)\alpha - (\alpha \cdot \alpha)\beta$, and use the inequality $\gamma \cdot \gamma \ge 0$.
3. The angle θ between α and β is defined to be
$$\theta = \arccos\left(\frac{\alpha \cdot \beta}{|\alpha| \, |\beta|}\right).$$
   In other words, $\alpha \cdot \beta = |\alpha| \, |\beta| \cos\theta$. Then α is perpendicular to β if and only if $\alpha \cdot \beta = 0$.
4. The cross product is anticommutative, that is, $\beta \times \alpha = -\alpha \times \beta$. In particular, $\alpha \times \alpha = \mathbf{0} = (0, 0, 0)$.
5. $|\alpha \times \beta|^2 + (\alpha \cdot \beta)^2 = |\alpha|^2 |\beta|^2$.

6. $\alpha \times \beta$ is perpendicular to each of its two factors. In fact, if n is of unit length and perpendicular to both α and β, then $\alpha \times \beta = \pm|\alpha|\,|\beta|\sin\theta\,n$. (Transpose one term in the preceding equation to the other side in order to conclude that $|\alpha \times \beta| = |\alpha|\,|\beta|\sin\theta$.)

Problem 1.8. We can identify the real number a with the element $(a, 0, 0, 0)$ in four-dimensional space. And we can identify the vector $\alpha = (a_1, a_2, a_3)$ with the element $(0, a_1, a_2, a_3)$ in four-dimensional space. In that way, we can think of a general quaternion $A = a + \alpha = (a, a_1, a_2, a_3)$ as a formal sum of a number and a vector. Adding two quaternions $A = a + \alpha$ and $B = b + \beta$ is trivial: $A + B = (a+b) + (\alpha+\beta)$. Multiplying them is less trivial. It took Hamilton some time to work out the proper rules for multiplying elements of four-dimensional spaces. (As we mentioned, Gibbs' work, which we are using to introduce this topic, actually came later.) The proper definition turns out to be $AB = (ab - \alpha \cdot \beta) + (a\beta + b\alpha + \alpha \times \beta)$. Notice that AB is in general different from BA, since the cross product is antisymmetric.

Show that 1, identified with the quaternion $\mathbf{1} = (1, 0, 0, 0)$, has the property $1A = A1 = A$ for all quaternions A.

Problem 1.9. Although the order of multiplication makes a difference for quaternions, they do resemble complex numbers in many ways. Quaternions have a real part and a vector part, whereas complex numbers have a real part and an imaginary part. The vector part of a quaternion behaves something like an imaginary number, since if $a = 0$, you find that $A^2 = (0+\alpha)(0+\alpha) = (-|\alpha|^2) + \mathbf{0}$, which is identified with the negative real number $-|\alpha|^2$. In other words, each vector α can be regarded as the square root of the negative of the square of its length.

Show that real numbers commute with all quaternions. That is, the real number a, identified with the quaternion $A_0 = a + \mathbf{0}$, has the property that $A_0 B = B A_0$ for all quaternions B.

Problem 1.10. Like a complex number, the quaternion $A = a + \alpha$ has a "conjugate" $\bar{A} = a - \alpha$. Show that $A\bar{A} = a^2 + |\alpha|^2 = a^2 + a_1^2 + a_2^2 + a_3^2$. We shall write $|A| = \sqrt{A\bar{A}} = \sqrt{a^2 + a_1^2 + a_2^2 + a_3^2}$. Since real numbers commute with all quaternions, it makes sense to define the reciprocal of the quaternion A as

$$\frac{1}{A} = \frac{1}{|A|^2}\bar{A}.$$

Notice that the quotient B/A isn't well defined. This symbol could mean either $(1/A)B$ or $B(1/A)$, and these two quaternions are in general not the same. Let $A = (1, 0, 0, 2)$ and $B = (0, 0, 3, 0)$. What are the two possible interpretations of B/A?

Problem 1.11. Describe several different ways of solving the plank-cutting problem of Example 1.1.

Question 1.1. Here are some questions of practical use in everyday life. Which of them pose arithmetic problems, and which pose algebra problems?

1. Looking at a stack of current household bills to be paid, how much money must you have in the bank or on hand in order to pay them?
2. With 60% of your grade in a course already determined and your current average at 85%, what average must you maintain for the remaining 40% of the course to ensure a semester average of 90%?
3. The distance s (in meters) that an object falls in t seconds, starting from rest and neglecting air resistance, is given by the formula $s = 4.9t^2$. How far will an object fall in 7 seconds?
4. Still referring to the formula $s = 4.9t^2$, how long will it take an object to fall 120 meters?

Question 1.2. Why is there no field having six elements?

6. Further reading

H. Behnke, F. Bachmann, K. Fladt, and W. Süss, *Fundamentals of Mathematics*, Vol. 1, translated by S. H. Gould, The MIT Press, Cambridge, MA, 1974.

Richard Courant and Herbert Robbins, *What is Mathematics?*, (second edition, revised by Ian Stewart), Oxford University Press, New York, 1996.

Lars Gårding, *Encounter with Mathematics*, Springer-Verlag, New York, 1977.

Albert C. Lewis, "Complex numbers and vector algebra," in *Companion Encyclopedia of the History and Philosophy of the Mathematical Sciences*, I. Grattan-Guinness, ed., Vol. 1, Routledge, London, 1994.

Equations and Their Solutions

In the previous lesson, we defined algebra as the study of equations and methods of solving them, and an equation as a problem in which the "input" consists of certain given numbers that have been produced by performing specified arithmetic operations on unknown numbers. The "output" of an algebra problem (its solution) consists of all possible values of the unknown numbers. In the present lesson, we shall do a preliminary analysis of the problem posed by an equation.

1. Polynomial equations, coefficients, and roots

The main object of interest in elementary algebra books consists of polynomial equations like the sixth-degree equation exhibited in the previous lesson, whose solutions are the six values of $\sqrt{2} + \sqrt[3]{2}$. Polynomial equations of degrees 2 and higher usually have more than one solution, as we can see by forming equations with given roots. For example, the simplest quadratic equation with roots u and v is $(x - u)(x - v) = 0$, which expands to $x^2 - (u + v)x + uv = 0$. Of course, it wouldn't be written this way. You would not see the coefficients *displayed* as $u + v$ and uv, and the whole equation might be multiplied by some constant a. What you would see would be $ax^2 + bx + c = 0$. Solving this one quadratic equation is equivalent to solving the *two* equations $u + v = -(b/a)$ and $uv = c/a$. Similarly, solving a cubic equation

$$ax^3 + bx^2 + cx + d = 0$$

is equivalent to solving the three equations

$$u + v + w = -\frac{b}{a},$$
$$uv + vw + wu = \frac{c}{a},$$
$$uvw = -\frac{d}{a}.$$

From this perspective, a cubic equation represents a *single* condition that must be satisfied by each of the roots of a *system* of three equations in three unknown numbers, and the variable x stands for any one of the roots u, v, w of that system. In that context, the polynomial equation looks like a secondary problem arrived at in the course of solving a more basic one. You can generalize from this case to an equation of degree n. It will have (in general) n solutions, and the sum of all products of those solutions taken k

at a time will be the coefficient of x^{n-k} in the equation, multiplied by $(-1)^k$ and divided by the coefficient of x^n.

The system of equations represented by a polynomial equation is of a very special type, as you can see. The roots occur *symmetrically* in each of its equations. There is nothing to prevent us from considering general systems of equations, in which the roots need not occur symmetrically, for example, $u + 3v = 5$, $uv = -10$. But in the end, if we were to solve these equations, we would reduce them to the symmetric case. For example, we could solve the first equation for u and substitute that result in the second. Thus, $u = 5 - 3v$, and $-10 = uv = (5 - 3v)v$, so that $3v^2 - 5v - 10 = 0$. This last equation is again a polynomial equation, and its roots v, w satisfy the symmetric system

$$
\begin{aligned}
v + w &= 5/3\,, \\
vw &= -10/3\,.
\end{aligned}
$$

Thus, it appears that symmetry lies at the heart of the problem of solving equations, even those that do not appear at the outset to have symmetry. As we shall see, in order to solve any higher-degree equation, it is necessary to *break* the symmetry somehow, and express some nonsymmetric function of the roots in terms of the coefficients, which are symmetric. Not to give away too much too soon, at this point we'll just say that the key to breaking the symmetry is extracting a root of a complex number.

The uses of this symmetry-breaking principle go beyond the problem of finding the roots of polynomials. The mathematical technique developed for exploring symmetry (group theory) turned out to be not only a central subject in modern mathematics but also a tool of incalculable value for modern physics and chemistry.

1.1. Geometric interpretations.

A geometric perspective can help us to understand the nature of an equation or system of equations. Using the quadratic equation $ax^2 + bx + c = 0$ as an example, we can think of the two solutions u and v as a point (u, v) in a two-dimensional space. Each coefficient represents a restriction on that point; that is, it represents a curve that the point is confined to, so that instead of having two "degrees of freedom," it has only one. Actually, as you may know, the equation $u + v = -b/a$ represents a line, and $uv = c/a$ represents a hyperbola. These two conditions are independent of each other, so that when we impose both simultaneously, the point has no freedom whatsoever to move around. The only freedom that remains is to interchange u and v, which really doesn't count, since we don't care which of the two variables represents which of the two numbers. This procedure is illustrated in Fig. 2, where we solve the equation $x^2 - 2x - 8 = 0$ by drawing the curves $u + v = 2$ and $uv = -8$ in a plane. Since u and v occur symmetrically, there is no need to label the axes. Either axis will serve equally well as the u or v axis. Each of the points

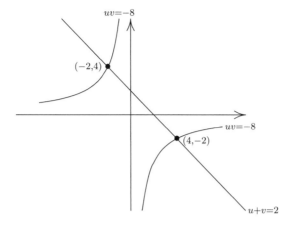

FIGURE 2. Solution of the equation $x^2 - 2x - 8 = 0$.

$(4, -2)$ and $(-2, 4)$ where the curves intersect represents the two possible solutions $x = 4$ and $x = -2$.

Remark 2.1. The line and hyperbola shown in Fig. 2 meet in two points. If the hyperbola lies in the other two quadrants of the plane (corresponding to the case when c/a is positive), the line may be tangent to it (corresponding to a double root of the quadratic equation), or even miss it entirely, since some quadratic equations have no solutions in real numbers. We can deal with that last possibility by working with complex numbers. However, Fig. 2 doesn't apply in the complex case. Two-dimensional complex space is actually four-dimensional space for our imagination, and so the situation is difficult to visualize and depict graphically.

2. The classification of equations

Many classifications of equations or systems of equations can be given, and the more one knows about algebra, the more complicated they become. Since we wish to keep things simple, our classification is the simplest possible. As far as we are concerned, there are only two kinds of systems: determinate and indeterminate. A *determinate* system is one that contains enough independent conditions (equations) to determine the solution uniquely. Strictly speaking, a polynomial equation of degree 2 or higher is not a determinate problem, since it has more than one solution. But, as we have shown, it is equivalent to a system of n equations in n unknowns that *is* determinate, since it has only one solution, except for permutations of the roots. For that reason, we shall classify polynomial equations as determinate problems.

In keeping with the idea that each new condition imposed reduces the number of degrees of freedom by one, we can expect that in general three independent conditions would suffice to determine three unknowns uniquely, except for certain symmetries, as noted above. In counting the number of equations, one must be careful to verify that they really are independent. For example, the three equations $x^2 + y^2 + z^2 = 16$, $x + y + z = 4$, and $xy + yz + zx = 0$ do *not* constitute three independent conditions, since the first equation can be deduced by squaring the second equation, multiplying the third by 2, and then subtracting. In this case, it is possible to determine two of the variables in terms of the third (say, y and z in terms of x), but one of them remains indeterminate.

It is possible to have more independent equations than unknowns. Such a system is determinate, but is said to be *overdetermined*. It may turn out to have no solutions at all. In order for an overdetermined system to have solutions, the equations must satisfy certain consistency conditions. We have really no use for overdetermined systems in these lessons and mention them only for the sake of completeness.

2.1. Diophantine equations. We shall also discuss indeterminate systems only briefly. Since an indeterminate system generally has a solution set that can be represented as a curve or surface in a higher-dimensional space, it requires methods from analysis (calculus) that can be omitted in a discussion of classical algebra. There is one important case, however, that should at least be mentioned. In order to reduce the solution set from a continuous curve or surface to a discrete set of points, mathematicians sometimes impose the additional requirement that the solutions be integers. An indeterminate system whose solutions are required to be integers is called a *Diophantine* system, after the (probably) third-century mathematician Diophantus of Alexandria (Egypt), who considered only positive rational solutions to his equations, since the positive rational numbers were the only numbers that he knew about. Some Diophantine equations have become quite famous. One that you are probably familiar with is the "Pythagorean" equation $m^2 + n^2 = p^2$, which has solutions $(3, 4, 5)$, $(5, 12, 13)$, $(15, 8, 17)$, and more generally, $(u^2 - v^2, 2uv, u^2 + v^2)$ for any integers u and v. Another famous example, much harder to solve, is "Pell's equation" $m^2 - Dn^2 = 1$, which has infinitely many solutions (m, n) for any positive integer value of D. (It has very little to do with the unfortunate John Pell, 1611–1685, after whom it is named.) In fact, the seventh-century Hindu mathematician Brahmagupta showed that if (a, b) is a solution, so is $(a^2 + Db^2, 2ab)$. This is because $(a^2 + Db^2)^2 - D(2ab)^2 = (a^2 - Db^2)^2 = 1$. On the other hand, the Diophantine equation $m^4 + n^4 = p^2$ has only the trivial solutions $m = 0$, $p = \pm n^2$ and $n = 0$, $p = \pm m^2$. We shall have only one occasion to refer to Diophantine equations in the lessons that follow (problems 11.1 and 11.2 in Lesson 11).

3. Numerical and formulaic approaches to equations

We are about to begin a general survey of the history of the portion of algebra that we have outlined, focusing on the study of determinate polynomial equations. The last bit of preparation we need involves a discussion of the meaning of the phrase "solving an equation."

3.1. The numerical approach. Since the solutions of equations are to be real or complex numbers, we can represent them, or their real and imaginary parts in the case of complex numbers, as potentially infinite decimal expansions. By solving an equation, do we mean simply finding better and better decimal approximations to a solution? In many cases, we do mean exactly that, and the development of fast and accurate ways to find approximate roots is still a legitimate way for a numerical analyst to spend time. We shall call this approach to equations the *numerical* approach. As just stated, the numerical approach focuses on the *result* of doing algebra, exhibiting the roots of the equation explicitly. If we write the expression $\sqrt{2}$, for example, we are merely using a concise way of specifying a real number whose square is 2, that is, a solution of the equation $x^2 - 2 = 0$. If you want to know "how big" $\sqrt{2}$ is, you need to look at decimal approximations to it. We need this potentially infinite process because we are accustomed to picturing points as strung out along a line representing the real numbers. If we omit that geometric interpretation, we can obtain a finite construction of $\sqrt{2}$, as shown in the previous lesson.

Teachers have often told their students that $\sqrt{2}$ is exact, while $1.41421\ldots$ is approximate. They have often insisted that only the "exact" value will do in mathematical work. Actually, $\sqrt{2}$ is an exact symbol only in the algebraic sense mentioned above. If $\sqrt{2}$ is to be interpreted geometrically, as the length of the diagonal of a square of side 1, and given a numerical value, we have to resort to some approximation. Doing so requires an algorithm for finding a sequence of rational numbers that approximate the real number $\sqrt{2}$. Rational numbers can be assumed to be known exactly, that is, with infinite precision, since their decimal expansions repeat after a finite period.

3.2. The formulaic approach. There is a second way of studying equations, going back to the idea introduced at the beginning of the previous lesson, that algebra attempts to reverse a sequence of arithmetic operations and move from the result of those operations to the data that were input. This approach is best explained using the example of the quadratic equation. Suppose that the equation $ax^2 + bx + c = 0$ has solutions u and v. From what was said above, we know that actually

$$ax^2 + bx + c = a(x - u)(x - v) = ax^2 - a(u + v)x + auv.$$

That is, $b = -a(u + v)$ and $c = auv$. As you may know, in this case it is possible to state explicitly, as a formula, exactly what needs to be done to solve the equation:

$$u = \frac{-(b/a) + \sqrt{(b/a)^2 - 4(c/a)}}{2}; \quad v = \frac{-(b/a) - \sqrt{(b/a)^2 - 4(c/a)}}{2}.$$

You can verify that, since $-(b/a) = u + v$ and $c/a = uv$, the expression under the square root sign in this formula is simply $(u - v)^2$, and hence the square root can be replaced by $u - v$. When you do that, you see that the two expressions for u and v are identities. The ancient Mesopotamians, some 3500 years ago, were the first people to make systematic use of what we now write as the "polarization" identity

$$\left(\frac{u + v}{2}\right)^2 = uv + \left(\frac{u - v}{2}\right)^2,$$

so called because it makes it possible to write the product uv as a difference of squares. They consistently used this relation implicitly (but wrote it out only with particular numbers in place of u and v, never in the abstract, as we have done here) to find two numbers given either their sum and product or their difference and product. In both cases, the roots appear symmetrically in the equation, but the polarization identity permits us to "break the symmetry" by taking the square root:

$$\frac{u - v}{2} = \pm\sqrt{\left(\frac{u + v}{2}\right)^2 - uv}.$$

In this way the quadratic equation $ax^2 + bx + c = 0$, which was the problem

$$u + v = -\frac{b}{a},$$
$$uv = \frac{c}{a},$$

is replaced by the linear system

$$\frac{u + v}{2} = -\frac{b}{2a},$$
$$\frac{u - v}{2} = \sqrt{\left(\frac{-b}{2a}\right)^2 - \frac{c}{a}},$$

and this last system is solvable. Thus, in a nutshell, the secret of solving a quadratic equation is to break its symmetry using the square root and the polarization identity, then replace it by a linear system whose solution is known.

Here we have a different approach to equations, in which we seek a *formula* that shows how to go from the coefficients to the roots. This approach focuses on the *relation between the coefficients and the roots*. Of course, if we have a formula, it should be possible to apply it and find numerical approximations, so the second approach in some sense contains the first. But the converse is not true. A numerical method of solving an equation does not necessarily tell us anything about the relation between coefficients and roots. We shall call this second approach to equations the *formulaic* approach. When it works, the solutions can be written as a finite expression involving the four operations of arithmetic together with root extractions. The central technique of the formulaic approach is the use of combinatorial methods—changing variables, rearranging terms, and the like—in order to obtain an equation of simpler form, whose solution may be obvious. We shall use both adjectives *formulaic* and *combinatorial* to describe this approach.

In the past, mathematicians in some nations such as China and Japan adopted the numerical approach, and they were very successful in applying it to equations of enormously high degrees. Other peoples, such as the medieval Muslims and modern Europeans, followed the formulaic route. The formulaic route greatly limits the degree of equations that can be studied in a practical manner. In fact, it is rarely practical to use even the formulas for solving equations of degrees 3 and 4, which were discovered in Italy during the sixteenth century. The formulaic approach did, however, lead to a magnificent intellectual edifice known as modern algebra, in particular the part of it known as *Galois theory*, which expresses the solvability of an equation by radicals in terms of a group of permutations.

As a byproduct of Galois theory, it became possible to solve some famous old problems from geometry. In the 1830s mathematicians using Galois theory were able to prove that no procedure using only a finite number of lines and circles drawn with straightedge and compass, starting from a line of unit length, can produce a square equal in area to a circle of unit radius, or an angle of 20° (the trisection of a 60° angle), or the side of a cube of volume 2. These constructions had been achieved by the ancient Greeks using curves more complicated than lines and circles, but here for the first time it was possible to state definitely that no solution could be found using only lines and circles. These "impossibility proofs" have unfortunately had no effect on the many eager amateurs, who, not understanding the problem that they imagine they are solving, construct hundreds of purported angle trisections and circle squarings every year.

4. Important concepts and principles in this lesson

If you have understood what was written in this lesson, you should have an adequate picture of the following concepts: determinate system, indeterminate system, overdetermined system, degrees of freedom, polynomial equation, Diophantine equation, numerical approach, formulaic approach, Galois theory, and impossibility proofs.

The main ideas contained in the present section are the following: (1) Solving a polynomial equation of degree n is equivalent to solving a system of n equations for n unknowns. (2) Solving an equation can be interpreted in two different ways. It may mean finding decimal approximations to the real and imaginary parts of the roots, or it may mean finding a formula that can be applied to the coefficients in order to express the solution.

5. Problems and questions

Problem 2.1. Sketch the curves $u+v = -(b/a)$, $uv = c/a$ for the following equations $ax^2 + bx + c = 0$. On the basis of the sketch, determine the number of real solutions the equation has and the approximate value of the roots.

$$3x^2 - 15x + 12 = 0$$
$$x^2 - 3x + 5 = 0$$
$$2x^2 + 10x + 12 = 0$$
$$3x^2 + 3x - 18 = 0$$

Problem 2.2. Solve the equation $x^2 + 2x + 2 = 0$ in the field with five elements. Does it have any solutions in the field with three elements? (Interpret the number 2 as -1 in that field.)

Problem 2.3. Which of the following two systems of three equations is determinate and which is indeterminate?

$$
\begin{array}{rcrcrcl}
x & + & 2y & - & 3z & = & 2, \\
2x & - & 3y & + & 4z & = & 1, \\
x & + & 9y & - & 13z & = & 5.
\end{array}
\qquad
\begin{array}{rcrcrcl}
x & + & y & + & z & = & 5, \\
x & + & 2y & + & 3z & = & 2, \\
x & + & 4y & + & 9z & = & 3.
\end{array}
$$

Problem 2.4. What condition on a, b, c, and d makes the following overdetermined system consistent?

$$
\begin{array}{rcrcrcl}
x & + & y & + & z & = & a, \\
x & - & y & + & z & = & b, \\
x & + & y & - & z & = & c, \\
x & - & y & - & z & = & d.
\end{array}
$$

Problem 2.5. Find (by guessing) a pair of positive integers m and n satisfying the Diophantine equation $n^m = m^n + 1$.

Question 2.1. Use the geometry of the situation, as illustrated in Fig. 2, to explain why a quadratic equation of the form $x^2 + ax + b = 0$ always has precisely one positive and one negative solution if $b < 0$.

Question 2.2. What does Fig. 2 become for the equations $x^2 + 2ax + a^2 = 0$ and $x^2 - 2ax + a^2 = 0$?

6. Further reading

I. Grattan-Guinness and W. Ledermann, "Matrix theory," in *Companion Encyclopedia of the History and Philosophy of the Mathematical Sciences*, Vol. 1, I. Grattan-Guinness, ed., Routledge, London, 1994.

Eberhard Knobloch, "Determinants," in *Companion Encyclopedia of the History and Philosophy of the Mathematical Sciences*, Vol. 1, I. Grattan-Guinness, ed., Routledge, London, 1994.

Helena Pycior, "The philosophy of algebra," in *Companion Encyclopedia of the History and Philosophy of the Mathematical Sciences*, Vol. 1, I. Grattan-Guinness, ed., Routledge, London, 1994.

Where Algebra Comes From

The simplest way to find out where algebra comes from is to look at some examples from long ago. In this lesson, we shall look at five of the earliest problems that can be regarded as algebra, from different places around the world.

1. An Egyptian problem

(About 3700 years ago, from the *Rhind Mathematical Papyrus*, Problem 24.) *A quantity and the seventh part of it have 19 as a sum. What is the quantity?*

Here we are told that a quantity has been divided by seven and the seventh part of it has been added to the original quantity, yielding 19 as the result. You will probably not find this problem difficult to solve if you write down the corresponding equation. When the Rhind Papyrus was written, some 3700 years ago, this problem was considerably more difficult. The author proceeded by a kind of guided guessing known in mathematical circles as the *method of false position*. He noted that if the quantity had been 7, the result of these operations would have been 8. Therefore, he reasoned, we must divide our guess (7) by 8, to make the 8 disappear and then multiply the quotient by 19, to get the desired result. As we would say, he needed to scale 7 by a factor of 19/8 so that the answer would come out to 19 instead of 8. Part of the difficulty to the Egyptian scribe came from the fact that multiplication and division as we know them did not exist. Multiplication was performed by repeated doubling and adding. To multiply 13×17, for example, the scribe would double 17, getting 34 (2×17), then double again to get 68 (4×17), then once again, getting 136 (8×17). Then, since $13 = 1 + 4 + 8$, he would add the numbers corresponding to 1, 4, and 8, namely 17, 68, and 136, finally getting the product 221. To do what we call dividing, the scribe would multiply the divisor by various integers and, if necessary, "unit" fractions with numerator 1—the main exception was that $2/3$ was allowed—until a set of products was reached that added up to the dividend. The sum of the integers and unit fractions by which the divisor was multiplied provided what we call the quotient. Dividing 7 by 8 was one of the simpler problems; it was merely a matter of dividing by 2, then repeating this operation twice more. The results were successively $3 + \frac{1}{2}$, then $1 + \frac{1}{2} + \frac{1}{4}$, and finally $\frac{1}{2} + \frac{1}{4} + \frac{1}{8}$. This number was then multiplied by 19, which is $1 + 2 + 16$, and the product was written as $(\frac{1}{2} + \frac{1}{4} + \frac{1}{8}) + (1 + \frac{1}{2} + \frac{1}{4}) + (8 + 4 + 2)$, which was finally simplified to $16 + \frac{1}{2} + \frac{1}{8}$. The restriction to unit fractions often led the scribe into very messy computations, since the double of a unit fraction had to be expressed as two other unit fractions. For example, $\frac{2}{7}$ was expressed as $\frac{1}{4} + \frac{1}{28}$.

2. A Mesopotamian problem

(Iraq, about 3500 years ago, cuneiform tablet AO 8862, now in the Louvre, Paris.) *I multiplied length and width to obtain area. I added the amount by which length exceeds width to the area and obtained* 183. *The sum of the length and width is* 27. *What are the length, width, and area?*

As this problem shows, even the very oldest texts contain problems that can still be challenging today. Here we encounter a problem with two independent unknowns, length and width. Correspondingly, we are given two sets of operations that have been performed on them and the results. One is simple; their sum is 27. The other is much more complicated: their product plus their difference is 183. It is not at all obvious how one can work backward from this information in order to get the two numbers. Trying to do this without putting it in modern notation is an interesting challenge. You can verify that the answer might be length 15, width 12 or length 14, width 13. The tablet gives only the first of these as an answer.

The author was apparently guided by the geometrical relations shown in Fig. 3, starting with a rectangle and gluing onto it first a strip of width one and length equal to the difference of the length and width of the original rectangle. The rectangle with this strip adjoined will have area 183, according to the problem. Then, gluing another strip of width one and length equal to the width of the original rectangle, and finally a strip of width 1 and length equal to the length of the original rectangle results in a new rectangle with the same length as the original but width increased by 2. Its area (the product of its length and width) will be $183 + 27 = 210$, and the sum of its length and width will be 29. Thus the problem has been simplified. We now have a rectangle in which we know the area (numerically, the product of length and width) and the sum of length and width. As shown in Lesson 2, the problem of finding two unknown numbers given their sum and their product is exactly the same problem as finding the two solutions of a quadratic equation. We know the sum (29) and the product (210) of the two dimensions. Hence the enlarged rectangle must be 15×14, and so the original one must be either 13×14 or 15×12.

3. A Chinese problem

(Han Dynasty, about 1900 years ago, from the *Nine Chapters on the Mathematical Art*, Chapter 6.) *A fast walker goes* 100 *paces in the time required for a slower walker to go* 60 *paces. If the slower walker has a head start of* 100 *paces, how many paces will be required for the faster walker to overtake the slower?*

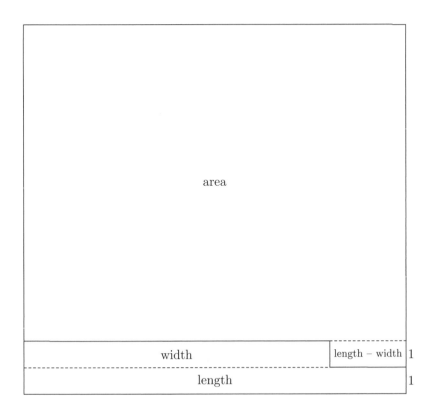

FIGURE 3. An ancient Mesopotamian algebra/geometry problem. Modified from *The History of Mathematics: A Brief Course*, second edition, John Wiley and Sons, New York, 2005, page 402.

You have very likely solved problems of this type already. Once the problem has been set up by taking the time required for the slower walker to go 60 paces as a unit, you see that the faster walker is gaining 40 paces per unit time. So you take as the unknown the number of units of time elapsed when the faster walker overtakes the slower. You know that 40 times this unknown number of units must equal 100 (the amount of the head start), and therefore the unknown must be $2\frac{1}{2}$. After that many units, the slower walker will have gone 150 paces and the faster one 250 paces.

4. An Arabic problem

(From the *Algebra* of Muhammed ibn-Musa al-Khwarizmi, about 1200 years ago.) *A man writes a will leaving 1* dirhem *cash and one-fifth of his estate to a friend, the rest to be divided equally between his two sons. When he dies, he leaves* 10 dirhems *cash on hand, and one of his sons owes him* 10 dirhems. *Find the amount of the* 10-dirhem *debt that must be added to the* 10 dirhems *cash on hand so that, when the estate is divided, the indebted son neither owes anything nor receives anything.*

The unknown in this problem is the portion of the debt that is to be added to the cash on hand to define the total estate, which will then be divided according to the will. The remainder of the debt was to be canceled, taken "off-budget," so to speak. Using the Arabic word for *thing* where we would use a single letter x, al-Khwarizmi reasoned that the estate consisted of $10 + x$ *dirhems*, where x is the amount that the indebted son would "pay" out of his share of the inheritance. The friend was entitled to $3 + x/5$, leaving $7 + 4x/5$ to be divided equally between the sons. Since each was to get $3\frac{1}{2} + 2x/5$, that is the amount x that the indebted son would have to pay. In other words, al-Khwarizmi derived the equation $x = 3\frac{1}{2} + 2x/5$, whose only solution is $x = 35/6$. This is the amount that the unindebted son receives, and the friend gets the remainder of the 10 *dirhems* cash on hand, that is, $25/6$ *dirhems*.

5. A Japanese problem

(Posed in 1670.) From 1600 to 1850 Japanese mathematicians published challenges to one another using area and volume problems. In one such problem, published in 1670 by Sawaguchi Kazuyuki (dates uncertain), two equal circles are tangent to each other, and each is tangent to a third circle whose diameter is five units larger than their diameters. All three are enclosed in a large circle, to which each is tangent. The area inside the large circle and outside the other three is 120. (See Fig. 4.) The problem asks for the diameters of the four circles. (Two of them are equal.)

This problem leads to one linear, one quadratic, and one cubic equation in the three diameters x, y, and z, namely

$$
\begin{aligned}
x + 5 &= y, \\
2\pi x^2 + \pi y^2 + 480 &= \pi z^2, \\
4y^2 z + 2xyz + xy^2 + xz^2 &= 4yz^2,
\end{aligned}
$$

where the letters are as shown in Fig. 4.

These equations have six solutions, but the only one that makes geometric sense is approximately $x = 7.58688$, $y = 12.58688$, and $z = 20.648$ respectively for the diameters of the smallest two circles, the larger circle, and the enclosing circle.

The first equation here obviously gives y in terms of x. When $x + 5$ is substituted for y in the second equation, we get $z^2 = 3x^2 + 10x + 25 + 480/\pi$. This value of z^2 can then be substituted into the third equation, and that equation can then be solved for z in terms of x:

$$
z = \frac{4\pi x^3 + 40\pi x^2 + (125\pi + 720)x + (250\pi + 4800)}{\pi(3x^2 + 25x + 50)}.
$$

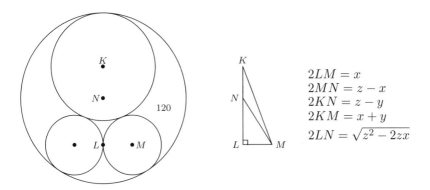

FIGURE 4. A Japanese algebra/geometry problem from 1670. Modified from *The History of Mathematics: A Brief Course*, second edition, John Wiley and Sons, New York, 2005, page 418.

Thus each value of x determines the values of y and z. If we square this last value of z, compare the result with the earlier value of z^2, and clear out the denominator, the result is an equation of degree 6 in x:

$$11x^6 + 220x^5 + \left(1900 - \frac{1440}{\pi}\right)x^4 + \left(8500 - \frac{24000}{\pi}\right)x^3 +$$

$$\left(20000 - \frac{120000}{\pi} - \frac{518400}{\pi^2}\right)x^2 + \left(25000 - \frac{360000}{\pi} - \frac{6912000}{\pi^2}\right)x$$

$$- \frac{1200000}{\pi} - \frac{23040000}{\pi^2} = 0\,.$$

This equation has six roots, four of which are complex and one of which is negative: $-9.673 - 2.49732i$, $-9.673 + 2.49732i$, $-1.66764 - 7.55092i$, $-1.66764 + 7.55092i$, and -6.21813. Here again, $i = \sqrt{-1}$. All of these numbers are only approximate, since the decimal expansion of π is infinite. Only the one positive root 7.58688 makes sense in the geometric problem.

6. Problems and questions

Problem 3.1. How many ways can you think of to solve a quadratic equation such as $x^2 - x + 1 = 0$ in a finite field, such as the field of three elements?

Problem 3.2. The quadratic formula for solving $ax^2 + bx + c = 0$ is

$$x = \frac{-b \pm \sqrt{b^2 - 4ac}}{2a}\,.$$

"Translate" this formula into the field with three elements (replace 4 by 1 and 2 by -1) and into the field with five elements. Under what circumstances will you be able to take the square root in this field? Is the translated formula still valid?

Question 3.1. How does each of these five examples fit the definition of algebra as finding unknown numbers given the result of performing operations on them? In each case, what are the operations performed?

Question 3.2. Which of the solutions described above are numerical and which are formulaic?

Question 3.3. Can the quadratic formula be translated into the field of two elements?

7. Further reading

Walter Eugene Clark, ed., *The* Aryabhatiya *of Aryabhata*, University of Chicago Press, Chicago, 1930.

Henry Thomas Colebrooke, *Algebra, with Arithmetic and Mensuration from the Sanscrit of Brahmegupta and Bhascara*, J. Murray, London, 1817.

John N. Crossley and Alan S. Henry, "Thus spake al-Khwārizmī: A translation of the text of Cambridge University Library Ms. ii.vi.5," *Historia Mathematica*, **17** (1990), 103–131.

Tobias Dantzig, *Number, the Language of Science*, Fourth edition, The Free Press, New York, 1967.

Richard J. Gillings, *Mathematics in the Time of the Pharaohs*, MIT Press, Cambridge, MA, 1972.

Lancelot Hogben, *Mathematics for the Million* (third edition), W. W. Norton, New York, 1952. (Hogben erroneously calls the *Aryabhatiya* of the fifth-century Hindu astronomer Aryabhata I by the name *Lilavati*, which is the name of a work by the twelfth-century mathematician Bhaskara II.)

Annick M. Horiuchi, "The development of algebraic methods and problem-solving in Japan in the late seventeenth and early eighteenth centuries," in *Proceedings of the International Congress of Mathematicians, Kyoto, Japan, 1990*, The Mathematical Society of Japan, 1991.

Lay-Yong Lam, "Jiu Zhang Suanshu (Nine Chapters on the Mathematical Art): An overview," *Archive for History of Exact Sciences*, **47** (1994) 1–51.

Yoshio Mikami, *The Development of Mathematics in China and Japan*, Chelsea, New York, 1961 (reprint of 1913 edition).

Otto Neugebauer, *The Exact Sciences in Antiquity*, Princeton University Press, Princeton, NJ, 1952.

Eleanor Robson, *Mesopotamian Mathematics, 2100–1600 BC: Technical Constants in Bureaucracy and Education*, Clarendon Press, Oxford, 1999.

Frederic Rosen, *The Algebra of Mohammed ben Musa*, Oriental Translation Fund, London, 1831.

V. S. Varadarajan, *Algebra in Ancient and Modern Times*, American Mathematical Society, Providence, RI, 1998.

B. L. van der Waerden, *Science Awakening*, Wiley, New York, 1963.

B. L. van der Waerden, *A History of Algebra from al-Khwārizmī to Emmy Noether*, Springer-Verlag, New York, 1985.

Yan Li and Shiran Du, *Chinese Mathematics: A Concise History*, translated by John N. Crossley and Anthony W.-C. Lun, Clarendon Press, Oxford, 1987.

Why Algebra Is Important

The title of the present lesson should perhaps be phrased as a question: *Is* algebra important? The examples presented in the previous lesson may not strike you as particularly practical. On the basis of those examples the only answer one could give to the question "Why did people solve these problems?" is "Because they could." If you delve into some early treatises on algebra, you may be even more discouraged in your attempt to feel the respect for the subject that the curriculum seems to require.

Consider, for example, the following problem from the *Lilavati* of the Hindu mathematician Bhaskara II (about 850 years ago):

> One pair out of a flock of geese remained sporting in the water, and saw seven times the half of the square root of the flock proceeding to the shore, tired of the diversion. Tell me, dear girl, the number of the flock.

You may be forgiven for thinking that it is not worth the trouble to write down the equation for the number of geese in the flock, that is, $x - 2 = \frac{7}{2}\sqrt{x}$, and solve it to get $x = 16$.

Or, consider the following problem from a treatise by Girolamo Cardano published about 450 years ago:

> The profit made by a certain business equals the cube of one-tenth of its capital. If the profit had been three ducats greater, it would have been exactly equal to the capital. What was the capital and what was the profit?

Again, rather than writing the equations $P = (C/10)^3$ and $P + 3 = C$ for profit and capital, then eliminating P between them to get $C^3 + 3000 = 1000C$, to find that the capital was 30 ducats and the profit 27 ducats, you might prefer simply to *ask the proprietors* how much their capital and profit were.

If you have gone very far in algebra, you have probably encountered problems of equal uselessness. One perennial favorite, which you may have seen, asserts that a man is now three times as old as his son and 10 years from now will be twice as old as his son and asks you to find their current ages. You must think to yourself, "Under what circumstances (outside of an algebra book) would I ever come to know these facts about the two without knowing their ages?"

As these examples show, algebra books have always found it difficult to produce motivating examples. Students may be excused for thinking that algebra is useful mostly in constructing puzzles for idle amusement. Even our glance into the actual treatises that have been written on algebra has not revealed any serious, practical purpose for solving equations of degree higher than the first.

Nevertheless, even if mathematicians developed algebra purely for amusement, the way people create and solve crossword puzzles, we nowadays have many reasons to thank them. As it became more subtle, algebraic reasoning acquired the compact notation of letters and symbols that we associate with the subject today. This process began very early in India, but picked up speed noticeably in the early seventeenth century in the work of François Viète (1540–1603) and René Descartes (1596–1650). By the time of Leonhard Euler (1707–1783), a century after Descartes, mathematical notation in Europe was nearly standardized in its present form.

Along with the use of letters came the notion of a *variable*, a symbol (usually a letter) representing an unspecified number. In applications, variables were used to denote quantities measured in specified ways. This new notation made it possible to rewrite the laws of physics in a much more compact way. Consider, for example, the way Johannes Kepler (1571–1630) originally stated his third law of planetary motion, in his 1619 treatise *The Harmonies of the World*, Book 5, Chapter 3: "The ratio between the periods of any two planets is the ratio of the $\frac{3}{2}$th power of their mean distances from the Sun." We would nowadays write this in other ways. Let T_1 and T_2 be the periods of two planets (length of a year on those planets) and r_1 and r_2 their mean distances from the Sun. Then

$$\frac{T_1}{T_2} = \left(\frac{r_1}{r_2}\right)^{3/2},$$

or, more typically,

$$\frac{T_1^2}{T_2^2} = \frac{r_1^3}{r_2^3}.$$

We remark in passing that the mean distance from the Sun is the average of the greatest and least distances. By Kepler's first law, a planet moves in an elliptical orbit with the Sun at one focus (on the major axis), and this mean distance is half of the major axis of the ellipse. For simplicity in what follows, we are going to consider only circular orbits, for which the distance r is constant.

The last equation makes it easy to take in at a glance what the relation between period and distance is, something that the prose statement given by Kepler does not do.

Time, distance, area, volume, mass, density, pressure, temperature, charge, current, energy, and many other important physical concepts are variables, and there are relations among those variables that can be expressed by equations, such as the ideal gas equation $PV = nRT$, or the

Stefan–Boltzmann law $E = \sigma T^4$ or Ohm's law $V = IR$ or Newton's law of gravitation $F = G_0(Mm/r^2)$, or the most famous of them all, Einstein's equation $E = mc^2$.

If that were the only advantage of algebraic notation, it would never have achieved the prominent place it now occupies in science. After all, Roman numerals can be used to record numerical data; and, although everyone knows what they mean, few would regard them as an effective way of acquiring new information. Would you care to divide DCCXLI by CCXLVII, for example, without converting to Hindu–Arabic numerals?

Algebra provided more than just a compact notation for writing down relations among variables. Its rules made it possible to manipulate those laws on paper and derive some of them from others. For example, a consequence of Kepler's third law is that the ratio T^2/r^3 of the square of a planet's period to the cube of its distance from the Sun is the same for all planets. In the exercises at the end of this lesson, you will be invited to use algebra to demonstrate that Kepler's third law and Newton's law of gravitation are equivalent statements, given certain basic facts of mechanics.

Even though the importance of algebra is proved beyond any doubt by its applications to the laws of physics, its usefulness would still be very limited, if not for the enrichment provided by the use of infinitesimal methods—the calculus. Differential calculus allows physical laws to be stated as relations between variables and their relative rates of change. Such relations are called *differential equations*, and it turns out that most of the important laws of nature have to be stated in precisely this way, as differential equations. Among the famous differential equations of physics are the heat equation, the wave equation, and the Schrödinger equation. Integral calculus provides a set of rules by means of which it is sometimes possible to eliminate the rates of change from a differential equation and replace it with an ordinary algebraic, exponential, or trigonometric relation between variables. In this way, physics has achieved prodigies of understanding about the universe. We shall now explore one of these more complicated examples where calculus is involved, leaving some simpler examples involving only algebra for the reader to explore in the exercises at the end of the lesson.

1. Example: An ideal pendulum

The example we are about to study must be accompanied by a warning that it involves some basic mechanics, a differential equation, and a function from advanced analysis, all of which the reader is to accept on trust, as raw facts. The reader is not expected to be able to fill in the details of the physics and mathematics below or to know what is meant by the Jacobi amplitude function that will be mentioned. (It is named in honor of Carl Gustav Jacobi (1804–1851), who introduced it in connection with elliptic functions.) Instead, our aim is to show how algebra and calculus can generate predictions for the behavior of an observable physical system on the basis of very general

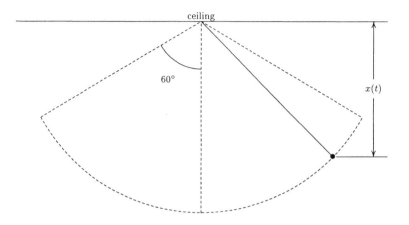

FIGURE 5. A pendulum swinging through a 120° arc.

principles written down in the language of algebra and analyzed by algebraic manipulation.

The point to be understood is that it is possible to use the laws of Newtonian mechanics to write down a differential equation, and after that equation is solved, it is possible to draw a graph of the variation of an observable quantity over time. The steps in between are beyond the scope of this book. For us, the mere fact that they are logically connected, so that the beginning determines the end, is the important thing. Although this point has already been made, it bears repeating that the really important applications of algebra come only after the study of calculus.

Imagine a cord or a rigid rod suspended from the ceiling of a room, free to pivot from its point of suspension, with a weight at the bottom of it. The cord or rod will be vertical when at rest. But you can pull it to one side, causing it to rise toward the ceiling. If you then let go, it will begin to oscillate as a pendulum. How can you describe this oscillation? One way to do so is to let the letter $x(t)$ denote the vertical distance from the ceiling to the weight at time t, as shown in Fig. 5.

Since the pendulum is below the ceiling, we'll assume $x(t)$ is a negative number. Physicists and mathematicians use the symbol $x'(t)$ to represent the rate at which $x(t)$ is increasing at time t. (A negative value of $x'(t)$ means $x(t)$ is decreasing.) From Newton's laws and a bit of integral calculus, one can derive the following law of oscillation:

$$(x'(t))^2 = 2g(x(0) - x(t))\left(1 - \left(\frac{x(t)}{L}\right)^2\right),$$

where g is the acceleration of gravity (9.8 meters per second per second) and L is the length of the cord or rod. This last equation relates the rate

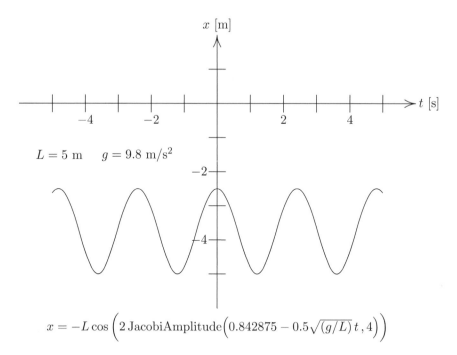

$$x = -L\cos\left(2\,\text{JacobiAmplitude}\left(0.842875 - 0.5\sqrt{(g/L)}\,t\,,4\right)\right)$$

FIGURE 6. Motion of a pendulum.

of increase of a quantity to the quantity itself. Such an equation, as mentioned above, is called a *differential equation*. From it, again using integral calculus, one can express $x(t)$ explicitly in terms of t. In the case where the oscillation begins with the pendulum making a 60° angle with the vertical, the expression is

$$(1) \qquad x(t) = -L\cos\left(2\,\text{JacobiAmplitude}\left(0.842875 - 0.5\sqrt{\frac{g}{L}}\,t\,,4\right)\right).$$

For the particular case when the pendulum has length $L = 5$ meters, the up-and-down oscillation can be graphed as shown in Fig. 6. The complete period of oscillation is just under 5 seconds (4.8165 seconds, to be more precise) as the pendulum moves from its maximum height of 2.5 meters below the ceiling to its minimum height of 5 meters below the ceiling, continues on to the same maximum at the opposite end of its swing, and then swings back again. Here we have a *prediction* that can be tested by constructing a pendulum 5 meters long and timing its swings from an initial angular displacement of 60°.

Turning to a general pendulum swinging through 120° arcs, we see by Eq. 1 that this model predicts that the period of oscillation is proportional to the square root of the length L. This prediction also can be tested by building pendulums of different lengths. If we could go to the Moon or some other heavenly body having a stronger or weaker gravitational field, we could also test the prediction that the period is inversely proportional to the square root of the acceleration of gravity. Thus, merely being able to read the language of algebra enables us to conjecture many possibilities about the world that can be checked by observation and serve as tests of the physical theories on which they are based.

2. Problems and questions

We chose to analyze a complicated illustration of the usefulness of algebra in the text above. To fix those ideas better, we now present the reader with a simpler example to work out. In the first four problems below, you will be given a set of dots to connect in order to "discover" the inverse-square law of gravitational attraction and test its validity from known data. For simplicity, we assume that all planetary orbits are circles.

Problem 4.1. We begin with the first attempt to analyze motion more complicated than simple motion at constant velocity. For constant-velocity motion, the well-known law is $s = vt$, where s is the distance, v the speed (rate), and t the time of the motion.

    The next step up is to consider *uniformly accelerated motion* in which the speed is proportional to the time, that is, $v = at$, where a is a constant called the *acceleration*.

Caution: You cannot combine this equation with the previous equation and deduce that $s = at^2$, since the two equations are not both valid for the same motion. The first one applies *only* when v is constant, and the second asserts that v is *not* constant.

    Algebra comes to our rescue here, aided by analytic geometry. The geometric relation "area = height × width," which applies to a rectangle (in a vertical plane), is of exactly the same form as the relation "distance = speed × time." Thus, we could represent the distance as the area below the curve that gives speed in terms of time. The two representations, for constant velocity and constant acceleration, are shown in Fig. 7. The area of the shaded region is numerically equal to the distance s.

    If the argument from analogy seems uncertain, it can be strengthened by introducing a bit of "infinitesimal" reasoning. Over a very short interval of time, the speed will be practically constant, and the constant-speed law will apply approximately. Divide the time interval from 0 to t into n equal pieces, that is, the pieces from $(k-1)t/n$ to kt/n, $k = 1, 2, \ldots, n$, as shown in Fig. 7, where we took $n = 8$. The distance traveled in that interval will be approximately $vt/n = akt^2/n^2$. More precisely, it will be at least $a(k-1)t^2/n^2$ and at most akt^2/n^2, that is, larger than the area of the shorter rectangle over that interval, and smaller than the area of the taller

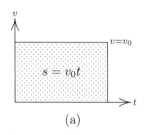
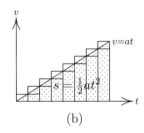

(a)   (b)

FIGURE 7. (a) Distance traveled at constant speed ($v = v_0$) is represented as the area of a rectangle, that is, $s = v_0 t$. (b) The Merton rule, when speed is directly proportional to time ($v = at$) and distance is represented as the area of the shaded triangle, so that $s = \frac{1}{2}at^2$.

rectangle. As a result, the total distance s traveled will satisfy

$$\frac{at^2}{n^2} + 2\frac{at^2}{n^2} + \cdots + (n-1)\frac{at^2}{n^2} < s < \frac{at^2}{n^2} + 2\frac{at^2}{n^2} + \cdots + n\frac{at^2}{n^2}.$$

These inequalities look cleaner when written as

$$\left(1 + 2 + \cdots + (n-1)\right)\frac{at^2}{n^2} < s < \left(1 + 2 + \cdots + n\right)\frac{at^2}{n^2}.$$

Here is your first simple algebra problem. Use the well-known identity

$$1 + 2 + \cdots + p = \frac{p(p+1)}{2}$$

to deduce that

$$\left(\frac{1}{2} - \frac{1}{2n}\right)at^2 < s < \left(\frac{1}{2} + \frac{1}{2n}\right)at^2.$$

Since n may be as large as desired, conclude that the law of uniformly accelerated motion is $s = \frac{1}{2}at^2$. *Hint:* If $s < \frac{1}{2}at^2$, then $s < (1/2 - 1/2n)at^2$ for some n, which is a contradiction, and likewise if $s > \frac{1}{2}at^2$.

This law is a version of what is known as the *Merton rule*, after Merton College, Oxford, where this rule was first stated in the thirteenth century. It is illustrated in Fig. 7.

Four centuries after this rule was first formulated, Galileo argued that a body falling near the Earth's surface has approximately constant acceleration. For such a body, the acceleration due to the Earth's gravitational attraction as we now say, is denoted by g, which has the value 9.8 meters per second per second. That is, during each second, the speed of the body increases by 9.8 meters per second. After 5 seconds, its speed will be 49 meters per second. (This approximation neglects certain things, chiefly air resistance to the falling body.)

Problem 4.2. The law of inertia, stated by Descartes in the seventeenth century, says that any unaccelerated motion will be motion in a straight line at constant speed. Now in nature, besides straight-line motion at constant

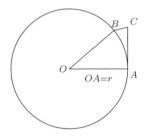

FIGURE 8. Forces on a body in circular motion.

speed, there are also many examples of motion in a circle at constant speed. To a very good approximation, for example, planetary orbits can be regarded as such a uniform circular motion.

Since uniform circular motion is not in a straight line, it must be accelerated. Obviously the acceleration is not constant. In fact, it is always directed toward the center, a direction that changes as the object moves. However, symmetry considerations show that the *magnitude* of the acceleration must be constant. What is that magnitude?

To find the answer, consider Fig. 8. A body in uniform circular motion on a circle of radius r at speed v will travel from A to B in a given time, whereas if it had moved without any acceleration, it would have moved from A to C in the same time. Thus, the acceleration that keeps it on the circle has caused it to "fall" from C to B. Trigonometry (with angles measured in radians) reveals that the distance s that it "falls" has a ratio to the square of the time fallen that is given by

$$\frac{s}{t^2} = \frac{1}{2}\frac{v^2}{r}\sqrt{\left(\frac{\sin(vt/2r)}{vt/2r}\right)^4 + \frac{4r^2}{(vt)^2}\left(1 - \frac{\sin(vt/r)}{vt/r}\right)^2}.$$

Now we regard this physical process of falling as taking place *continuously*, so that the time interval t here is "infinitely small." From calculus it is known that for very small angles $\sin(\theta)/\theta \approx 1$ and $1 - \sin(\theta)/\theta \approx \theta^2/6$.

It follows that the quantity under the square root is approximately 1 when t is a very short interval of time. That means that if s/t^2 is replaced by the constant $\frac{1}{2}v^2/r$, the *relative* error is very small. It tends to zero as t tends to zero. Therein lies the whole secret of using calculus in physics. If the *relative* error tends to zero, it can be treated as zero on the infinitesimal level.

After these long preliminaries, you get a short task to perform: By comparing this formula with the Merton rule $s/t^2 = \frac{1}{2}a$, show that the magnitude of the "instantaneous" acceleration must be v^2/r.

Problem 4.3. Constant linear acceleration for falling bodies on Earth was generally accepted after Galileo's work, giving the law relating distance and time as $s = \frac{1}{2}gt^2$.

Uniform circular motion appeared to be of a different, celestial order. To take the simplest example, the Moon revolves around the Earth in an approximate circle, with a (sidereal) period of 27.3 days. (As you know, the time from one full Moon to the next is closer to 29.5 days. However, in that time the Moon must actually traverse about 390°, since the Earth moves about 30° around the Sun in the same period. A full 360° trip around the Earth takes the Moon only 27.3 days.)

In the seventeenth century, people began to speculate that the same "g" force that causes bodies to fall to Earth might also be responsible for the acceleration that holds the Moon in its orbit. It was not expected that g would be as large at the distance of the Moon's orbit as it is at the surface of the Earth. After all, the Moon is about 60 times farther from the Earth's center than is the Earth's surface. More precisely, the ratio is about 60.2687. If the Moon's acceleration really is due to g, how big must g be at that distance?

It turns out that g at the radius of the Moon's orbit is 0.00272327 meters per second per second. To see why, note that $g = v^2/r$, where r is the radius of the Moon's orbit. You can do the computation yourself. Just observe that the Moon travels a distance equal to $2\pi r$ in 27.322 days. From that information you can compute v. There are 86,400 seconds in a day, and the average radius of the Moon's orbit is $r = 3.844 \times 10^8$ meters.

How does this value compare with the value of g on the Earth's surface, which is 9.8 meters per second at the equator and 9.86 meters per second at the North Pole? The ratio is

$$\frac{9.8}{0.00272327} = 3598.62.$$

This is remarkably close to 3600. Since the radius of the Earth is 6.3781×10^6 meters, the ratio that the distance from the Moon to the Earth's center bears to the Earth's radius is $(3.844 \times 10^8)/(6.3781 \times 10^6) \approx 60.2687$, which is close to 60. It thus appears that increasing the distance by a factor of 60 decreases the gravitational constant by a factor of 3600, which is 60^2.

Thus, the hypothesis that the gravitational acceleration of a body decreases in proportion to the square of the distance seems reasonable. It is reasonable on geometric grounds, if you think of this attractive force spreading out evenly, with the same total amount on all spheres about the center of attraction. Since the areas of spheres are proportional to the squares of their radii, the intensity (attraction per unit area) must be inversely proportional to that area.

How could we test such a conjecture on a larger scale? Newton pondered this problem, and asked how the orbital period would have to decrease with increasing distance from the center of attraction. The orbital periods of the planets had been known since ancient times, and their distances from the Sun, once the Copernican theory has been accepted, are also easy to compute from observation. Your problem is to compute the relation between the orbital period T and the radius r of a planet's orbit, *assuming* that the gravitational attraction of the Sun produces an acceleration that is inversely proportional to r^2. In other words, assume that $v^2/r = C/r^2$ (where C is a constant whose value we don't need to know) and eliminate v from this relation by using the relation $v = 2\pi r/T$. If you do this, you will have derived *Kepler's third law*, which was derived by Kepler as the best fit to observational data a quarter-century before Newton was born. Newton realized in 1665 that Kepler's third law and the inverse-square law linked the two kinds of acceleration and that this coincidence strongly suggested that the acceleration of a planet is gravitational in nature. However, not having accurate data on the size of the Earth and the radius of the Moon's orbit, he didn't get the kind of close agreement that we have obtained when he tested the hypothesis. He put the computation aside for a few years, until improved geographic studies provided better agreement.

Problem 4.4. Show that the assumption that the gravitational acceleration due to a central body is v^2/r, together with Kepler's third law, implies that the gravitational acceleration must decrease according to the square of the distance. (This is the converse of what was done in the last problem. In other words, the inverse-square law of attraction and Kepler's third law are *equivalent* statements for circular orbits, given that the acceleration of a body in uniform circular motion is v^2/r.)

Problem 4.5. As the preceding problems show, algebraic relations among physical variables can give insight into the universe, but only if they agree with physical measurements. The close agreement of the inverse-square law with the actual acceleration of the Moon is an excellent example.

Here is another example, in which agreement with observation strongly suggests a physical principle. The principle consists of three parts:

1. *Coulomb's law.* The repulsive force F between two like point charges q_1 and q_2 at distance r is

$$F = \frac{1}{4\pi\varepsilon} \frac{q_1 q_2}{r^2},$$

   where ε is the *dielectric permittivity* of the medium in which the charges are found. For "empty" space this permittivity is experimentally found to be

$$\varepsilon_0 = 8.84 \times 10^{-12} \text{ coulomb}^2/\text{newton-m}^2.$$

2. The notion of *magnetic induction.* A magnetic induction of magnitude B exerts a force F on a charge q moving with speed v at an angle θ with the direction of the magnetic induction, where

$$F = Bvq\sin\theta,$$

The physical units of B are force/(charge×velocity), in other words, newtons times seconds divided by coulombs times meters. One newton-second per coulomb-meter is called a *weber*, after Wilhelm Weber (1804–1891). An intriguing aspect of electromagnetic theory is that a moving electric charge produces a magnetic induction at a point P, according to the equation

$$B(P) = \mu \frac{vq}{r^2},$$

where r is the distance from the charge q to point P and μ is another physical constant called the *magnetic permeability* of the medium. For "empty" space, the permeability is

$$\mu_0 = 4\pi \times 10^{-7} \text{ weber-meter-second per coulomb}.$$

Eliminating the weber, we find

$$\mu_0 = 4\pi \times 10^{-7} \text{ newton-second}^2 \text{ per coulomb}^2.$$

Notice that the product of the two fundamental constants ε_0 and μ_0 has the dimensions of seconds-squared per meter-squared, in other words, $\varepsilon_0\mu_0$ represents the square of the reciprocal of a velocity. Your task is to compute the numerical value of the velocity $1/\sqrt{\mu_0\varepsilon_0}$ in meters per second.

3. Electromagnetic theory predicts that an oscillating electric field and an oscillating magnetic field can propagate as a coupled wave, each generating the other, provided they propagate at the velocity $1/\sqrt{\mu_0\varepsilon_0}$.

If you have done your part of this problem, you know that this velocity is precisely the velocity of light! This point was noted as early as 1856, and the natural conclusion was drawn explicitly by James Clerk Maxwell (1832–1879) in 1861: Light, which the human race has always perceived directly because of its action on our eyes, is actually an epiphenomenon resulting from the interaction of electric and magnetic fields. This insight produces an amazing simplification and unification of our knowledge of the universe.

Question 4.1. Summarize the arguments given in these five problems, and explain how exactly algebra helps to relate physical phenomena to one another. Which of the steps here would have been meaningless without algebraic notation?

Question 4.2. Algebra also serves as a guide to physical reasoning in connection with electromagnetic theory. According to Newtonian mechanics, two observers moving with constant velocity relative to each other must agree on all the forces that they observe. However, they do not agree about electric fields that they observe. If one observes an electric field intensity E and a magnetic induction B, a second observer moving with velocity v relative to the first will observe fields E' and B' given by

$$\begin{aligned}
E' &= E + v \times B, \\
B' &= B.
\end{aligned}$$

Thus the two observers agree about the total force exerted by these two fields, but they disagree about the intensity of the electric field, even as they agree about the magnetic induction. This strange asymmetry disappears if we reconstruct electromagnetic theory within the context of special relativity. In that case, we find

$$E' = \alpha(E + v \times B) + \frac{(1 - \alpha)E \cdot v}{v \cdot v}v,$$

$$B' = \alpha\left(B - \frac{1}{c^2}v \times E\right) + \frac{(1 - \alpha)B \cdot v}{v \cdot v}v,$$

where $\alpha = 1/\sqrt{1 - v \cdot v/c^2}$. Notice that if $c = \infty$, then $\alpha = 1$, and the relativistic law becomes the classical law.

To what extent is the greater symmetry that we can read in the algebraic formulas from relativity a clue that relativity theory is a better explanation than Newtonian mechanics?

3. Further reading

Ivor Grattan-Guinness, *The Fontana History of the Mathematical Sciences*, Fontana Press, London, 1997.

Morris Kline, *Mathematics and the Physical World*, Crowell, New York, 1959.

Henri Poincaré, *Science and Method*, translated by Francis Maitland, Barnes & Noble Books, New York, 2004.

Numerical Solution of Equations

We now turn to the main purpose of classical algebra, the development of methods of solving polynomial equations. Recall that there are two interpretations of the problem of solving an equation, leading to two different approaches to its solution. In the present lesson, we discuss the numerical approach. Although numerical methods are important in finding both real and complex solutions, the essential ideas can be presented without the use of complex numbers. For that reason, we shall take advantage of this simplicity and discuss only methods of finding real solutions of equations with real coefficients. With this restriction some equations, such as $x^2 + 1 = 0$, will not have any solutions at all, but that problem will be addressed in our next lesson.

1. A simple but crude method

Let us try to invent a numerical method of solving an equation, using just our intuition. As an example, consider the polynomial $p(x) = x^3 - 5x^2 + 10x - 5$. We want to solve the equation $p(x) = 0$. Let us begin by computing the values of $p(x)$ at small integers. Who knows? We may get lucky and hit a value at which $p(x)$ equals 0. With very little effort, we find $p(0) = -5$, $p(1) = 1$. Although we didn't find a root, we can see that there must be one somewhere between 0 and 1, since the values of $p(x)$ change from negative to positive as x increases from 0 to 1. We might even hazard a guess that the root is closer to 1 than to 0, since the value at 1 is closer to zero than the value at 0 is. What we can be sure of is that there is a root in the interval $(0, 1)$.

Our procedure from now on is to bisect the interval in which the root is confined. It is possible that the value of $p(x)$ at the midpoint will be exactly 0, in which case the process stops. If not, its sign will be opposite to the sign of the value at one of the endpoints, and we can take the half of the interval for which the values are of different signs at the two endpoints as a new interval. In this way, we either find a root or cut its range of possible values in half at each stage. As you probably know, repeatedly discarding half of object reduces its size fairly quickly. A hand calculator may help in the process. Here are the first few steps.

Step 1: $p(0.5) = -1.125$. We now have the root confined to the interval $(0.5, 1)$, whose midpoint is 0.75.

Step 2: $p(0.75) = 0.109375$. We now have the root confined to the interval $(0.5, 0.75)$, whose midpoint is 0.625.

Step 3: $p(0.625) = -0.458984$. We now have the root confined to the interval $(0.625, 0.75)$, whose midpoint is 0.6875.

You can see how to continue this process. If we were to stop at this point and offer the new midpoint 0.6875 as the root, we could be sure we were making an error no larger than 0.0625. Good numerical methods give the root as 0.724318, so that in fact our error would be about 0.037. The "exact" value of the root, as will be explained in Lesson 7, is one you probably don't wish to contemplate, namely

$$\frac{1}{3}\left(5 + \sqrt[3]{\frac{50}{13 - 3\sqrt{21}}} - \sqrt[3]{\frac{5}{2}(13 - 3\sqrt{21})}\right).$$

Modern mathematicians and computer scientists have worked out some amazingly efficient, accurate, and rapid algorithms for finding numerical approximations to roots. It is not our purpose to discuss any of these methods in detail. Instead, we want to look at numerical methods as a primary approach to solving equations. For that purpose, we can confine ourselves to the Chinese culture, in which these methods were highly developed. The Chinese method has one important feature in common with ours. It works by finding smaller and smaller intervals in which the root must lie, and finds them by keeping the values of the polynomial different at the two endpoints. But where our method cuts the size of the interval in half at each step, the Chinese method finds the next decimal digit at each step, in other words, cuts the interval down to one-tenth of its previous size. To do that, it must be more sophisticated than the method we have presented.

2. Ancient Chinese methods of calculating

Perhaps the Chinese developed numerical methods to such a high degree because of their numbering system and the fact that they used mechanical methods of calculation, in the form of tally sticks, counting boards, and eventually the abacus. Let us first discuss the Chinese numbering system.

Unlike the ancient Egyptians and Greeks, who had special symbols—the Greeks used their alphabet—for 1, ...,9, 10, ..., 90, 100, ..., 900, the Chinese had only the symbols for 1, ..., 9, 10, 100, 1000, and so on. The important theoretical difference between these two systems is that, in the Chinese system, the numbers 20, ..., 90, 200, ..., 900 had to be written using the multiplicative principle, so that 300 was written using the symbol 3 followed by the symbol for 100. Since Chinese symbols *are* words, we can get a feeling for this notation by writing the number 3845 as 3 thousand, 8 hundred, 4 ten, 5. The system uses 10 as a base, but it is not strictly a place-value system, since the power of 10 represented by a symbol actually accompanies the symbol. To go from this system to a purely place-value system, one needs mainly a symbol for an empty place (zero) and then the rather obvious realization that the symbols for the powers of ten need not be written down, since they can be inferred from physical location. This step was not taken in China, perhaps because of the advanced mechanical methods used. The rows or columns on a counting board can represent different powers of 10, and where we would put a zero, one could simply leave that row or column empty. Hence there was no immediate need to pass to a purely place-value written system with a zero.

But arithmetic's loss in this case is algebra's gain, since it was realized early on that the rows and columns on a counting board could be used to represent different objects in a problem or different powers of an unknown quantity. We shall say just a few words about the use of columns to represent different objects before taking up our main theme, the solution of polynomial equations using counting-board methods.

2.1. A linear problem in three unknowns. The fundamental early Chinese treatise on mathematics, the *Nine Chapters on the Mathematical Art*, contains a problem involving three varieties of wheat, of varying quality. Two bundles of the first kind plus one bundle of the second kind, when threshed out, will produce one bushel of grain, as will three bundles of the second kind plus one bundle of the third kind and four bundles of the third kind plus one bundle of the first kind. The problem is to determine how many bushels of grain are in one bundle of each kind of wheat. If these quantities are x, y, and z, then the conditions of the problem give

$$\begin{aligned} 2x + y &= 1, \\ 3y + z &= 1, \\ x + 4z &= 1. \end{aligned}$$

Since these are linear equations, one can always solve them by successive elimination. However, in modern linear algebra, a more efficient method has been developed that mimics to some extent the Chinese counting board. The coefficients of the system are arranged in a rectangular array called a *matrix*, in this case consisting of three rows and four columns:

$$\begin{pmatrix} 2 & 1 & 0 & 1 \\ 0 & 3 & 1 & 1 \\ 1 & 0 & 4 & 1 \end{pmatrix}.$$

On the Chinese counting board, the zeros would correspond to empty squares. The matrix is not merely a convenient, static way of writing down the equations. It can be used dynamically by manipulating its rows and columns. If we were to do this, we would probably interchange the first and last equations, and then proceed as follows:

$$\begin{pmatrix} 1 & 0 & 4 & 1 \\ 0 & 3 & 1 & 1 \\ 2 & 1 & 0 & 1 \end{pmatrix} \longrightarrow \begin{pmatrix} 1 & 0 & 4 & 1 \\ 0 & 3 & 1 & 1 \\ 0 & 1 & -8 & -1 \end{pmatrix} \longrightarrow$$

$$\longrightarrow \begin{pmatrix} 1 & 0 & 4 & 1 \\ 0 & 1 & -8 & -1 \\ 0 & 3 & 1 & 1 \end{pmatrix} \longrightarrow \begin{pmatrix} 1 & 0 & 4 & 1 \\ 0 & 1 & -8 & -1 \\ 0 & 0 & 25 & 4 \end{pmatrix}.$$

To get from the first of these to the second, subtract the first row from the third row twice, replacing the third row with the result each time. To get from the second to the third, interchange the second and third rows. To get from the third to the fourth, subtract the second row from the third row three times, again replacing the third row with the result each time.

The last row in this final array is interpreted as the equation $25z = 4$, so that $z = \frac{4}{25}$. The second row represents the equation $y - 8z = -1$, and since we know the value of z, we get $y = 8z - 1 = \frac{7}{25}$. Finally, the first row represents the equation $x + 4z = 1$, so that $x = 1 - 4z = \frac{9}{25}$.

Our point in introducing this example is that the manipulations that we performed on this matrix are well adapted for performance using counters (sticks or pebbles) placed on the squares of a counting board.

3. Systems of linear equations

In the lessons that follow, we shall have occasion to discuss systems of linear equations in connection with polynomial equations. A slight digression is needed at this point to establish some important facts about such systems. To keep things simple, we shall take a system of three equations in three unknowns as typical. We need only two simple facts. First, whether a linear system

$$\begin{aligned} a_{11}u + a_{12}v + a_{13}w &= q \\ a_{21}u + a_{22}v + a_{23}w &= r \\ a_{31}u + a_{32}v + a_{33}w &= s \end{aligned}$$

can be solved or not, depends on a single number, called the *determinant* of the matrix of coefficients. Without going into details, we merely remark that the determinant is

$$a_{11}a_{22}a_{33} + a_{12}a_{23}a_{31} + a_{13}a_{21}a_{32} - a_{11}a_{23}a_{32} - a_{13}a_{22}a_{31} - a_{12}a_{21}a_{33}.$$

If this equation looks like a tangled mess of subscripts, notice that the first subscripts in each term are all in their natural order $1, 2, 3$. The second subscripts are permutations of that ordering, with a plus sign if the permutation is even and a minus sign if it is odd. (The notion of even and odd permutations will be explained in Lesson 9.) If the determinant is nonzero, the system always has one and only one solution for u, v, and w.

Second, a case of special interest occurs when the coefficients a_{ij} form what is called a *Vandermonde matrix*, after Alexandre Vandermonde (1735–1796). In a Vandermonde matrix the entry in row i and column j is $a_{ij} = a_j^{i-1}$, that is, in the 3×3 case such a matrix has the form

$$\begin{pmatrix} 1 & 1 & 1 \\ a_1 & a_2 & a_3 \\ a_1^2 & a_2^2 & a_3^2 \end{pmatrix}.$$

Its determinant is $(a_1 - a_2)(a_2 - a_3)(a_3 - a_1)$, and hence nonzero if the numbers are all distinct. This case will be important in several of the following lessons.

4. Polynomial equations

We shall now show that these same counting-board techniques can be used to solve a higher-degree equation. The usefulness of a matrix arrangement in solving a polynomial equation is that different rows can be used to represent different powers of the unknown. As an example let us consider the quadratic equation $x^2 - 124x - 917 = 0$. If $p(x) = x^2 - 124x - 917$, we can see that $p(100) = 10000 - 12400 - 917 = -3317 < 0$ while $p(200) = 40000 - 24800 - 917 = 14283 > 0$. Thus there is a root between 100 and 200, and so the first digit of the root is 1.

To get the second digit, we take 100 as a "base value" and let $x = 100 + y$, where now we know that $0 < y < 100$. We need to rewrite the equation in terms of y. The Chinese found a very simple way to do this on a counting board, by filling in the blanks in the following array:

$$\begin{array}{cccc} 1 & 1 & 1 & 1 \\ -124 & & & 0 \\ -917 & & 0 & 0 \end{array}.$$

Before giving the rule for completing this array, we note two things: (1) each entry in the top row is equal to the leading coefficient of the equation, while the left-hand column is simply the full set of coefficients; and (2) the zeros here would be merely empty squares on the counting board. We inserted them as "stop signs" for the procedure about to be described, but they have an additional advantage that will appear shortly.

The rule for filling in the array is simple. Work from left to right and top to bottom. To find what goes in an empty space, multiply the entry immediately above the space by the current "base value" (100) and add the adjacent number on the left. The result is

$$\begin{array}{cccc} 1 & 1 & 1 & 1 \\ -124 & -24 & 76 & 0 \\ -917 & -3317 & 0 & 0 \end{array}.$$

The coefficients of the equation that y has to satisfy can now be read diagonally downward from right to left, that is, $p_1(y) = y^2 + 76y - 3317 = 0$. We will not take the time to explain in full why this procedure always works, although it is not difficult to analyze. You can verify that it has given the correct result in this case, since $0 = p(x) = (y+100)^2 - 124(y+100) - 917 = y^2 + 200y + 10000 - 124y - 12400 - 917 = y^2 + 76y - 3317$.

Now we know that $p_1(y)$ has a zero between 0 and 100. Calculation shows that $p_1(30) = -137 < 0$ and $p_1(40) = 1323 > 0$. Hence the zero is between 30 and 40, and so the second digit of the root is 3.

To get the third digit, we repeat the process, writing $y = 30 + z$ (using 30 as the current "base value") and filling in the array to get

$$
\begin{array}{cccc}
1 & 1 & 1 & 1 \\
76 & 106 & 136 & 0 . \\
-3317 & -137 & 0 & 0
\end{array}
$$

Thus z satisfies $p_2(z) = z^2 + 136z - 137 = 0$, and we know that z is between 0 and 10. We then find very quickly that $z = 1$ gives an exact root, so that $x = 131$ is the root of the original polynomial.

Although the equation is now solved, we might continue to experiment with this method. What would happen if we continued, letting $z = 1 + w$? What would the equation for w look like? The method would yield

$$
\begin{array}{cccc}
1 & 1 & 1 & 1 \\
136 & 137 & 138 & 0 . \\
-137 & 0 & 0 & 0
\end{array}
$$

In other words, w would satisfy $p_3(w) = w^2 + 138w = 0$, so that $w(w+138) = 0$. What this tells us is that w might be *either* 0 (which we already knew) *or* -138, so that z might have been either 1 or -137, y might have been either 31 or -107, and x might have been either 131 (as we found) or -7. You can verify that $x = -7$ is indeed a solution of the original equation $x^2 - 124x - 917 = 0$.

4.1. Noninteger solutions. Before considering cubic equations, we need to work one more example of this procedure to introduce a small complication that arises when the solutions are not integers. We illustrate it by finding the zeros of the polynomial $p(x) = 28x^2 - 23x - 15$. We start as usual by noting that $p(1) = -10$ and $p(2) = 51$, so that there is a root between 1 and 2. As before, we let $x = 1 + y$ and get the equation for y from the array

$$
\begin{array}{cccc}
28 & 28 & 28 & 28 \\
-23 & 5 & 33 & 0 . \\
-15 & -10 & 0 & 0
\end{array}
$$

Thus, y satisfies $p_1(y) = 28y^2 + 33y - 10 = 0$, and y is between 0 and 1. Since we want the next digit of the solution, we should try the numbers 0.1, 0.2, 0.3, and so on, as values of y until we find the point where $p_1(y)$ changes sign. It is simpler, however, to do a decimal shift and consider $10y$ instead of y. That is, we let $z = 10y$, so $y = z/10$. It is quite simple to see that z satisfies $q_1(z) = 28z^2 + 330z - 1000 = 0$, and this is easy to remember, since all we have to do is adjoin the zeros already in the array to the coefficients.

By trial, we find that $q_1(2) = -228 < 0$ and $q_1(3) = 242 > 0$, so the next digit will be 2. We then write $z = 2 + u$ and continue.

Again, since u is between 0 and 1, it is simpler to multiply it by 10 and write $v = 10u$, $u = v/10$. The array

$$
\begin{array}{cccc}
28 & 28 & 28 & 28 \\
330 & 386 & 442 & 0 \\
-1000 & -228 & 0 & 0
\end{array}
$$

tells us that v satisfies $q_2(v) = 28v^2 + 4420v - 22800 = 0$ and that v is between 0 and 10. This time, we find that $v = 5$ gives an exact solution. Therefore the solution of the equation is $x = 1.25$.

If we wanted to know the other solution, we could continue the procedure one more step, as we did above. The array would be

$$
\begin{array}{cccc}
28 & 28 & 28 & 28 \\
4420 & 4560 & 4700 & 0. \\
-22800 & 0 & 0 & 0
\end{array}
$$

In other words, if $v = 5 + w$, then w satisfies $28w^2 + 4700w = 0$, so $w = 0$ (as already found) or $w = -\frac{4700}{28} = -\frac{1175}{7}$. Then $x = 1 + y = 1 + z/10 = 1.2 + v/100 = 1.25 + w/100 = 5/4 - 1175/700 = -1200/2800 = -3/7$.

5. The cubic equation

To show that this procedure is perfectly general, we shall solve a cubic equation by the same method. To do this, we need one extra row and one extra column. The polynomial for which we shall find a zero is $p(z) = 4x^3 - 7x^2 + 7x - 3 = 0$. Since $p(0) = -3$ and $p(1) = 1$, there is a root between 0 and 1.

We then let $x = y/10$ and rewrite the equation in terms of y, that is, $q(y) = 4y^3 - 70y^2 + 700y - 3000 = 0$. By guessing or trial, we find that $q(7) = -158$ and $p_1(8) = 168$, and we see that there is a root of $q(y)$ between 7 and 8. We now let $z = 7 + y$.

The array that gives the equation for z is

$$
\begin{array}{ccccc}
4 & 4 & 4 & 4 & 4 \\
-70 & -42 & -14 & 14 & 0 \\
700 & 406 & 308 & 0 & 0. \\
-3000 & -158 & 0 & 0 & 0
\end{array}
$$

We find that z must satisfy the equation $p_1(z) = 4z^3 + 14z^2 + 308z - 158 = 0$. We now write the equation for $w = 10z$, which is $q_1(w) = 4w^3 + 140w^2 + 30800w - 158000 = 0$. We know that w is between 0 and 10. Since $q_1(5) = 0$, we have now found the solution: $x = 0.75$.

The procedure described here generates the successive decimal digits of a real solution of any equation with real coefficients that has at least one real solution. If the solution has a finite decimal expression, the procedure terminates when it generates the solution exactly.

In this way (working with sufficient patience and accuracy), it is possible to find any number of decimal digits of a root of any equation with real coefficients, no matter its degree. The Japanese mathematician Seki Kowa (Seki Takakazu, 1642–1708) is said to have solved an equation of degree 1458, over a period of several days, on the floor of a large room ruled into squares. (This claim should be treated skeptically!)

In 1819, a technique essentially the same as this ancient Chinese method, except that it applied to infinite series as well as polynomials, was developed by the British scholar William George Horner (1787–1837). It was taught for about a century in American high-school algebra books under the name *Horner's method*, with the computations simplified using "synthetic division."

6. Problems and questions

Problem 5.1. Using the examples given above as a model, solve the equation $x^2 - 7 = 0$ to two decimal places. When you finish, you should have the first two digits of $\sqrt{7}$, truncated rather than rounded off. In other words, you should know that the root lies between 2.64 and 2.65. The computations should be easy, at least at the first stage, because of the absence of a linear term.

Problem 5.2. Find a two-place approximation to $\sqrt[3]{3}$ by solving the equation $x^3 - 3 = 0$.

Problem 5.3. Solve the equation $x^3 + x^2 + x + 1 = 0$ in the finite field of five elements using the Chinese method. Make your first guess $x = 1$, and verify that the new equation you get is indeed the result of substituting $x = 1 + y$ into this equation.

Question 5.1. We have been vague about the way to find the initial approximation to a root. Why *must* there be a root r of the polynomial $a_0 z^n + a_1 z^{n-1} + \cdots + a_{n-1} z + a_n$ satisfying

$$|r| < 1 + \frac{|a_1|}{|a_0|} + \frac{|a_2|}{|a_0|} + \cdots + \frac{|a_n|}{|a_0|}?$$

Question 5.2. Is it necessary to try each digit $0, 1, \ldots, 9$ in succession in order to find the successive digits of a a solution? If it were, why would we need the algorithm, since we could just keep substituting longer and longer decimal expansions in the original equation?

Question 5.3. Is the Chinese method more efficient than the bisection algorithm we developed in the first section?

Question 5.4. In any field, finite or infinite, how does the Chinese numerical procedure tell you that you have found a root precisely?

7. Further reading

Lay-Yong Lam and Tian-Se Ang, *Fleeting Footsteps. Tracing the Conception of Arithmetic and Algebra in Ancient China*, World Scientific, River Edge, NJ, 1992.

Ulrich Libbrecht, *Chinese Mathematics in the Thirteenth Century*, MIT Press, Cambridge, MA, 1973.

Jean-Claude Martzloff, "Li Shanlan (1811–1882) and Chinese traditional mathematics," *The Mathematical Intelligencer*, **14** (1982), 32–37.

Margaret McGuire, "Horner's Method," in *A Source Book in Mathematics*, David Eugene Smith, ed., Dover, New York, 1959.

Part 2

The Formulaic Approach to Equations

Lessons 6 and 7 are devoted to the earliest phase of the formulaic approach to solving equations. This phase is characterized by attempts to express the unknown quantity in the equation in terms of some other quantity that satisfies a simpler equation. By such manipulations, mathematicians succeeded early in solving quadratic equations. The solution of cubic equations took much longer and required a second layer of recombination of variables in order to produce the result. Two different solutions of the general cubic equation were obtained by such techniques, but cubic equations represented the limit of its direct applicability. To progress beyond the cubic, it was necessary to supplement this combinatorial technique with the concept of a resolvent.

LESSON 6

Combinatoric Solutions I: Quadratic Equations

In the preceding lesson, we discussed ways of finding the root of an equation by generating its decimal expansion. This approach leaves the relationship between the input (coefficients) and output (roots) of the problem obscure. In the present section, we study attempts to clarify that relation and find rules for getting from the input to the output.

1. Why not set up tables of solutions?

As a transition between the numerical approach we have just discussed and the formulaic approach we are about to study, let us consider another possible proposal for solving equations: Solve a large number of equations with different coefficients and record the solutions in a table, so that one could simply *look up* the solution.

At first glance, this proposal seems preposterous. If we wished to list all the different quadratic equations $ax^2 + bx + c = 0$ with coefficients a, b, and c indexed at intervals of say 0.01 for values between, say 0 and 10, we would have a billion entries in our table. Obviously, that wouldn't work. But with a little thought and requiring a little extra work on the part of the person who uses the table, we can reduce the size of the table considerably. First of all, if $a = 0$, this isn't a quadratic equation at all; and if $a \neq 0$, then we can get an equation having the same solutions by dividing out a, so that we get $x^2 + (b/a)x + c/a = 0$, which we shall write as $x^2 + px + q = 0$. If you want to know the solution of the original equation, for which you know a, b, and c, you would first compute $p = b/a$ and $q = c/a$, then look up the answer in a much shorter table. However, if p and q both ran from 0 to 10 in increments of 0.01, that table would still have a million entries, far too many.

At the expense of still more work for the table user, we could reduce the number of parameters in the table still further, from 2 to 1, by making the substitution $y = x + \frac{1}{2}p$, that is, $x = y - \frac{1}{2}p$. The equation $x^2 + px + q = 0$ then becomes $y^2 - py + \frac{1}{4}p^2 + py - \frac{1}{2}p^2 + q = 0$, which can be rewritten as $y^2 = N$, where $N = \frac{1}{4}p^2 - q$. Thus the table we construct would be a table of square roots, and letting N vary from 0 to 10 in increments of 0.01 would produce a table of only 1000 entries. Given our original equation $ax^2 + bx + c$, we'd first compute p and q, as before, then compute $N = \frac{1}{4}p^2 - q$, then do a lookup in the table to get y, and finally subtract $\frac{1}{2}p$ from y to get x. In fact, tables of square roots do exist. Before the advent of calculators and computers, the technique just described was the lazy way to solve quadratic equations numerically.

Notice that while attempting to construct a compact table for solving the quadratic equation, we have actually produced a formula for doing so:

$$x = y - \frac{1}{2}p = \sqrt{\frac{1}{4}p^2 - q} - \frac{1}{2}p = \sqrt{\frac{b^2}{4a^2} - \frac{c}{a}} - \frac{b}{2a}.$$

In fact, this will be precisely the quadratic formula that you have probably already learned if we put an ambiguous sign in front of the square root:

$$x = \frac{-b \pm \sqrt{b^2 - 4ac}}{2a}.$$

This formula tells us how to get from the data (a, b, c) to the two roots. Moreover, it tells us when a quadratic equation with real coefficients has no real roots. The *discriminant* $D_2 = b^2 - 4ac$ is negative in that case and only in that case. This quantity is called the *discriminant* because it discriminates between the cases of two distinct roots and one double root. When it is zero, the only root is $x = -b/(2a)$, but it is customary to count it as a double root since in that case $ax^2 + bx + c = a(x + b/(2a))^2$. As you can easily compute, $b^2 - 4ac$ has a simple expression in terms of the roots u and v of the equation. Since $b = -a(u + v)$ and $c = auv$, the discriminant is $a^2((u + v)^2 - 4uv) = a^2(u - v)^2$. Strictly speaking, algebraists define the discriminant to be merely $(u - v)^2$. We shall take this quantity as the strict definition of the discriminant, but also use the term more loosely and refer to $D_2 = b^2 - 4ac$ as the *quadratic discriminant*. The important fact is that, since the discriminant is symmetric in u and v, it can be expressed in terms of the elementary symmetric functions $u + v$ and uv; in other words, it can be computed from the coefficients of the polynomial, without having to find the roots. Also, its square root, which is an *algebraic* function of the coefficients, is a *rational* function of the roots. That is, expressed in terms of the coefficients, this square root requires a radical sign, but in terms of the roots, it is a polynomial.

The discriminant gives us some useful information. As mentioned above, when the discriminant is zero ($u = v$), the polynomial $ax^2 + bx + c$ is a perfect square, namely $(\sqrt{a}(x - u))^2$. Further, when $b^2 - 4ac$ is negative (with real values of a, b, and c), we get complex roots for the equation. This interpretation of a negative quadratic discriminant is not the obvious one, and in fact, *it was not the solution of quadratic equations that led to the creation of imaginary numbers*. For centuries, it was simpler just to say that some quadratic equations have no solutions. What eventually led mathematicians to accept imaginary numbers was the *cubic* equation, as we shall see in the next lesson.

We have now progressed from a numerical approach through the attempt to build a table of solutions and arrived at a formula. We now ask what the formula means. If we need to find a numerical value for x, then we have to be able to get a numerical value for the square root. *Is that possible?* There are two obvious objections. One is that the square root operation is not among the four basic operations of arithmetic. In order to solve quadratic equations, we need to use this operation. Do we have a numerical procedure for getting an approximation to the square root of a positive real number?

The answer to that question is an emphatic "Yes." One quick way of getting the square root of N is known as the Newton–Raphson approximation, which starts with any guess. For example, we could start with $x_0 = N$, which is a bad guess unless N is close to 1. If that guess is too large, N/x_0 will be too small, and vice versa, since the product of x_0 and N/x_0 is N. Therefore, if we average them, we will improve on at least one of these two guesses. The Newton–Raphson algorithm then proceeds by the following recursion:

$$x_{n+1} = \frac{1}{2}\left(x_n + \frac{N}{x_n}\right).$$

It can be shown that $\{x_n\}$ converges very rapidly to a square root of N. One could also use the Chinese method of solving the equation $y^2 - N = 0$, which gives the successive decimal digits of the square root of N.

With that difficulty taken care of, we still have to wonder what happens if $b^2 - 4ac < 0$. In that case, we must resort to complex numbers to get a square root. But if we are going to allow complex numbers to be roots, we should allow them to be coefficients as well. If a, b, and c are complex numbers, then $b^2 - 4ac$ is also, very likely, a complex number. Is it possible to take the square root of a complex number? How do we find a complex number $z = u + iv$ such that $z^2 = w$, where $w = r + is$? By writing the equations for the real and imaginary parts of the equation $z^2 = w$, that is,

$$
\begin{aligned}
u^2 - v^2 &= r, \\
2uv &= s,
\end{aligned}
$$

and solving these two quadratic equations, you will find that the complex number

$$z = \sqrt{\frac{\sqrt{r^2 + s^2} + r}{2}} + i\sqrt{\frac{\sqrt{r^2 + s^2} - r}{2}}$$

has the required property, provided the signs of the two square roots are chosen properly (equal if $s > 0$, opposite if $s < 0$). Here the quantity under the square root signs is a nonnegative real number for all real values of r and s, and so we have reduced taking the square root of a complex number to taking the square root of nonnegative real numbers.

It appears that our problem is solved. Having a numerical implementation of the square root for all possible numbers, real or complex, we can adjoin the square root operation to the set of four numerical operations previously allowed, and confidently say that we have a solution for any quadratic equation. Strictly speaking, we should be more cautious. When we say we can solve every quadratic equation, we mean *if* we can solve every "pure" quadratic equation $y^2 = N$. But since we are quite sure that we can solve these equations, we can now state that we have solved the problem of the quadratic equation completely.

2. The quadratic formula

In this section we begin a systematic, abstract study of formulas for solving polynomial equations, which we shall continue in Lesson 8 for the case of cubic equations. To that end and for later reference, we need some uniform notation. Knowing what we do about the relationship between coefficients and roots, we write a typical quadratic polynomial as $x^2 - ax + b$, where a is the sum of the two roots of the polynomial, and b is their product.

By a formula for finding the roots of a polynomial, we mean an algebraic expression $x(a, b)$ formed from a and b such that $(x(a, b))^2 - ax(a, b) + b \equiv 0$, that is, this function is *formally* zero. We already know such a formula:

$$x(a, b) = \frac{a}{2} + \sqrt{\frac{a^2}{4} - b}.$$

Any formula that supplies the roots of the general quadratic equation formally must be a double-valued formula, since in general a quadratic equation has two roots. Since a rational function is single-valued, it cannot generate all the roots. Therefore we must expect any quadratic formula to contain a square root. To argue another way, rational operations do not lead outside a field. Hence if there were a rational quadratic formula, every quadratic equation would be solvable without enlarging the smallest field containing the coefficients. That, we know, is impossible.

We can rewrite the quadratic formula as

$$x = x(a, b) = \frac{a}{2} + z,$$

where $z = \sqrt{a^2/4 - b}$. (Our reason for complicating things by introducing an extra letter z to stand for the square root will appear when we continue this study in the next lesson.) In order to get a genuine algebraic formula, we omitted the $\pm$ sign usually included in this formula. As it stands, the square root is ambiguous. When actual numbers are substituted for the variables a and b, the square root may represent either of two numbers, except in the unusual case when $b = a^2/4$.

Remark 6.1. Here and below, when we write a root $\sqrt[m]{R}$ or $R^{1/m}$, where R is an algebraic formula, it is understood that this expression may be assigned any of m different values when numbers are substituted for the parameters and variables that occur in the formula R. However, we do insist that if $\sqrt[m]{R}$ or $R^{1/m}$ occurs more than once in a formula, the *same* value must be chosen in all occurrences when numbers are substituted for letters. In general, there is no sense in which $\sqrt[m]{R}$ and $\sqrt[m]{S}$ are the "same" root if R and S are different algebraic formulas.

If we substitute the value of $x(a,b)$ for x in the polynomial $x^2 - ax + b$ and then replace z^2 by $a^2/4 - b$, the result will be

$$(a - a)z + \left(\frac{a^2}{4} - b + \frac{a^2}{4} - \frac{a^2}{2} + b\right).$$

But elementary algebra shows that both the coefficient of z and the term independent of z are formally, identically zero here as functions of a and b. As a consequence, each of the two functions that are possible interpretations of the square root leads to a function $x(a,b)$ that is identically zero when substituted into the polynomial, and hence will yield a root of the polynomial when *any* particular numerical values are substituted for a and b, *independently of the choice of the numerical square root.* In other words, the formula "works" for any particular numerical values of a and b, no matter how you choose the square root.

Notice something else: If the two roots of the polynomial are u and v, then $a = u + v$ and $b = uv$, so that $a^2/4 - b = (u - v)^2/4$, and $z = (u - v)/2$ or $(v - u)/2$. In this way z is seen to be a *polynomial* when expressed in terms of the roots, despite having a radical sign when written in terms of the coefficients. Indeed, it is $(\alpha_1 u + \alpha_2 v)/2$, where $\alpha_1 = 1$ and $\alpha_2 = -1$ are the two square roots of unity. Although this seems a pretentious way of writing this simple fact, we shall see that it generalizes nicely to the cubic equation as well.

It is a fact, not difficult to prove, that any formula that is symmetric in the roots u and v can also be written in terms of a and b. For example, $u^2 + v^2 = a^2 - 2b$, $u^3 + v^3 = a^3 - 3ab$, and so on. What we have discovered above is that, if root extractions are allowed, we can also express the *nonsymmetric* function $u - v$ in terms of the coefficients as well. That step is crucial in solving the equation. In fact, by writing $u - v$ in terms of the coefficients, we obtain a simple system of *linear* equations for u and v:

$$\begin{aligned} u + v &= a, \\ u - v &= \sqrt{a^2 - 4b}, \end{aligned}$$

which is equivalent to the system

$$\begin{aligned} u + v &= a, \\ uv &= b, \end{aligned}$$

that the equation $x^2 - ax + b = 0$ represents.

Moreover, this linear system always has a unique solution, since its matrix is a Vandermonde matrix.

3. Problems and questions

Problem 6.1. Solve the equation $x^2 - ix + (1 - i) = 0$, where $i = \sqrt{-1}$. Check your answer by direct substitution into the polynomial. Find the square root of the discriminant by using the formula given above.

Problem 6.2. Find a condition on the *complex* numbers a, b, and c that is necessary and sufficient for the equation $ax^2 + bx + c = 0$ to have *at least one* real root.

Problem 6.3. In the field of two elements, the quadratic equation $x^2 + x + 1 = 0$ has no solutions. Let us introduce a larger field in which it has two solutions u and v. From what we know of the relation between coefficients and roots, these must satisfy $u + v = -1 = 1$ and $uv = 1$. Therefore, in this larger four-element field, we must have the following tables for addition and multiplication:

+	0	1	u	v
0	0	1	u	v
1	1	0	v	u
u	u	v	0	1
v	v	u	1	0

+	0	1	u	v
0	0	0	0	0
1	0	1	u	v
u	0	u	v	1
v	0	v	1	u

Use the tables to verify that indeed $u^2 + u + 1 = 0$ and $v^2 + v + 1 = 0$. Which of the four elements of this field have square roots in the field?

Problem 6.4. For complex numbers u and v, we have seen that the system of two equations $u + v = -b/a$ and $uv = c/a$ is equivalent to the quadratic equation $ax^2 + bx + c = 0$ satisfied by u and v, and u and v are the only two complex numbers that satisfy this quadratic equation. Notice that this system of two equations is no longer symmetric if u and v are quaternions, since uv is in general different from vu. Does this asymmetry make any difference when it comes to solving a quadratic equation? Find *all* of the quaternions $X = x + \boldsymbol{\xi}$ that satisfy the quadratic equation $X^2 + r^2 = 0$, where r is a real number, identified with the quaternion $r + \mathbf{0}$. *Hint:* See Problem 1.9.

Question 6.1. Why is it impossible for a quadratic equation with real coefficients to have one real and one nonreal root?

Question 6.2. Can a finite field be algebraically closed?

4. Further reading

"Al-Khwārizmī. Quadratic equations," in *A Source Book in Mathematics 1200–1800*, D. J. Struik, ed., Princeton University Press, Princeton, NJ, 1986.

Vera Sanford, "Cardan's treatment of imaginary roots," in *A Source Book in Mathematics*, David Eugene Smith, ed., Dover, New York, 1959.

LESSON 7

Combinatoric Solutions II: Cubic Equations

The solution of a quadratic equation by formula is sufficiently sophisticated that few attempts were made to solve any equations of higher degree by formula during the early days of algebra. Hindu and early Arabic treatises go as far as quadratic equations in the solution of determinate problems, but no further. On the other hand, Chinese mathematicians were not troubled by cubics, since they solved them numerically, as we saw in Lesson 5. Later Arabic treatises by Umar al-Khayyam Sharaf al-Din al-Muzaffar al-Tusi contain most of what is needed to solve cubic equations graphically and by formula.

In the present lesson, we follow the approach used in the previous one, considering how we might construct a *table* of solutions for cubic equations. With luck, we might once again wind up with a *formula* for the solution and not actually have to construct the table. Obviously, we shall have to allow another operation, namely extracting cube roots. Since we have numerical ways of doing that, we shouldn't hesitate to do it with *real* numbers. As long as only real numbers are involved, we assume that cube roots can be taken; how to take the cube root of a *complex* number is a question that we shall consider when the situation arises. Taking the cube root of a positive number r gives the side of a cube that is r times as large as a cube of side 1. This is an ancient problem known as *doubling the cube* in the case when $r = 2$. The Greeks were able to solve the general problem of multiplying the volume of a cube by r using conic sections, specifically parabolas and hyperbolas. As we shall see below, conic sections do provide a *graphical* solution of general cubic equations.

1. Reduction from four parameters to one

Condensing the steps a bit, since we have already been over this ground, we note that we don't actually need a four-parameter table indexed by a, b, c, and d to solve every cubic equation $ax^3 + bx^2 + cx + d = 0$. First, we can divide out a if it is nonzero. (If it is zero, the equation isn't really a cubic equation.) Next, the substitution $x = y - b/3a$ will reduce the equation to the form

$$y^3 + py + q = 0,$$

where we could compute p and q if we needed to, knowing a, b, c, and d. Thus it appears that our table of solutions to the cubic equation can be a two-parameter table.

Can we do still better? In fact, we can. If we let $y = z\sqrt{p}$, we find that the equation becomes

$$p\sqrt{p}z^3 + p\sqrt{p}z + q = 0,$$

which we can rewrite as

$$z^3 + z = -\frac{q}{p\sqrt{p}}.$$

Thus we need to solve only this one type of cubic equation in order to solve them all. If we were to pursue this route and set up a table of solutions, we would provide "operating instructions" telling the user how to compute first p and q from the coefficients a, b, and c, then compute $-q/(p\sqrt{p})$, use this last number to look up z in the table, and then finally compute $x = z\sqrt{p} - b/3a$.

This equation, is not a pure equation, since it contains the term z. For that reason, we have not yet arrived at a formula for the solution. Still, we could construct a table of reasonable size. As a pseudohistorical note, tables have been found in Mesopotamia dating back several thousand years, giving the values of $w^3 + w^2$ for an indexed set of values of w from 1 to 30. The substitution $y = q/(pw)$ in the equation $y^3 + py + q = 0$ would have led us to the equation $w^3 + w^2 = -q^2/p^3$. Thus, these ancient Mesopotamian tables *could be used* to solve any cubic equation, if the table were made big enough. However, it would be very far-fetched historically to say that they were constructed or ever used for that purpose.

We have not yet been able to change the unknown in a cubic equation so as to get a "pure" equation $y^3 = N$, which could be solved by allowing the extraction of a cube root. This reduction is difficult, and much preliminary work by Muslim and medieval European mathematicians was needed before the appropriate substitution was discovered in the seventeenth century. When that step was finally taken, it came as something of an anticlimax, since a different formulaic solution of the general cubic had already been found in Italy a century earlier.

2. Graphical solutions of cubic equations

The Persian mathematician Umar al-Khayyam (Omar Khayyam, 1048–1131) extended the application of conic sections to the solution of cubic equations. The conic sections, which were the subject of a long treatise by Apollonius of Perga in the third century BCE, had been invented by the Greeks to solve two problems for which straight lines and circles had proved inadequate. These were the construction of a cube twice as large as a given cube (or, more generally, having a given ratio to a given cube), and the trisection of an arbitrary angle. We now know that these two problems, taken together, are equivalent to the problem of finding the cube root of a complex number. Umar al-Khayyam wrote a treatise on algebra showing how to solve any cubic equation using conic sections. He warned his readers in the preface that no one should undertake his work without having first mastered the early books of the treatise of Apollonius.

Since negative numbers were not yet recognized, Umar al-Khayyam had to distribute the terms on the two sides of an equation so that every coefficient was positive. Counting the possibility of missing terms (zero coefficients), one can form 14 different types of cubic equations. We shall illustrate just one of these using the equation

$$x^3 + 2x^2 + 9x = 45.$$

Umar al-Khayyam showed that the unique positive value of x satisfying this equation can be obtained as the x coordinate (as we would now say) of one of the two points of intersection of the following two curves:

$$xy = 15,$$
$$x^2 - 3x + y^2 - 6y = 1.$$

It is not difficult to work out how Umar al-Khayyam knew that these two equations would do the job. (See Problem 7.3.) The first of these equations is a hyperbola; the second represents a circle of radius 3.5 with center at $(1.5, 3)$. One of the two points of intersection is $(5, 3)$, but the value $x = 5$ does not satisfy the original cubic equation. The other point of intersection is approximately $(2.34505, 6.39645)$, and its x coordinate is the desired root. The solution is illustrated in Fig. 9. You can verify that solving the first equation for y ($y = 15/x$), substituting that value into the second equation, and multiplying by x^2 to clear out the denominators, leads to the equation

$$(x - 5)(x^3 + 2x^2 + 9x - 45) = 0.$$

Hence the x value at the point of intersection must be either 5 or a root of the original cubic equation. At the same time, the procedure shows that this way of breaking the cubic into two quadratics does not lead to any formula for solving the cubic. It *displays* the root as a line that could be measured, but does not express the length of that line as an explicit formula. As you will be able to prove at the end of the present lesson, the "exact" value of this x is

$$\frac{1}{3}\left(-2 - 23\sqrt[3]{\frac{2}{1361 + 9\sqrt{23469}}} + \sqrt[3]{\frac{1}{2}(1361 + 9\sqrt{23469})}\right).$$

3. Efforts to find a cubic formula

Despite the extensive work of Umar al-Khayyam, continued by Sharaf al-Din al-Muzaffar al-Tusi (ca. 1135–1213), a formulaic solution of the cubic equation eluded the Muslim mathematicians of the Middle Ages. When the knowledge that they had acquired passed into Europe, it was eagerly seized upon by scholars, and at last, in the early sixteenth century, several Italian mathematicians found formulas for solving cubic and quartic equations. The first honors go to Scipione del Ferro (1465–1525) of the University of Bologna, but more generality and understanding was achieved by Niccolò

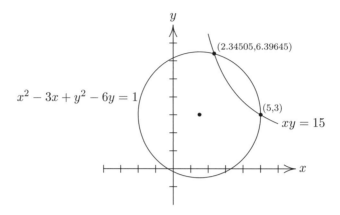

FIGURE 9. Umar al-Khayyam's solution of the cubic equation $x^3 + 2x^2 + 9x = 45$.

Tartaglia (1500–1557) of Brescia, Girolamo Cardano (1501–1576) of Padua and his student Ludovico Ferrari (1522–1565), and Rafael Bombelli (1526–1572).

The solution of the cubic depends first on the general reduction to the form $y^3 + py + q = 0$, which we have already given. This reduction dovetails nicely with the identity $(r - s)^3 + 3rs(r - s) + (s^3 - r^3) = 0$, which is true for all values of r and s. This identity shows that we could take $y = r - s$ as a solution if we could choose r and s so that $3rs = p$ and $s^3 - r^3 = q$. In the case when $p = 0$, the solution is trivial: $y = -\sqrt[3]{q}$. Hence we shall assume that $p \neq 0$, in which case r and s must also be nonzero. If we set $r = p/(3s)$ and substitute this value into the second equation, we get $s^3 - p^3/(27s^3) = q$, which can be rewritten as $s^6 - qs^3 - p^3/27 = 0$. Now the substitution $z = s^3$ gets us a quadratic equation: $z^2 - qz - p^3/27 = 0$, one of whose solutions is

$$s^3 = z = \frac{q + \sqrt{q^2 + 4p^3/27}}{2} = \frac{q}{2} + \sqrt{\frac{q^2}{4} + \frac{p^3}{27}}.$$

Then

$$r^3 = s^3 - q = -\frac{q}{2} + \sqrt{\frac{q^2}{4} + \frac{p^3}{27}}.$$

Hence we now have a *formula* for solving this cubic:

$$y = r - s = \sqrt[3]{-\frac{q}{2} + \sqrt{\frac{q^2}{4} + \frac{p^3}{27}}} + \sqrt[3]{-\frac{q}{2} - \sqrt{\frac{q^2}{4} + \frac{p^3}{27}}}.$$

This formula is called *Cardano's formula* for solving the cubic. It does actually work—sometimes. For example, consider the equation $y^3 + 60y -$

$992 = 0$, that is, $p = 60$, $q = -992$. The formula yields

$$y = \sqrt[3]{496 + \sqrt{496^2 + 20^3}} + \sqrt[3]{496 - \sqrt{496^2 + 20^3}} =$$
$$= \sqrt[3]{496 + 504} + \sqrt[3]{496 - 504} = \sqrt[3]{1000} + \sqrt[3]{-8} = 10 - 2 = 8.$$

At other times, while it works, it gives the answer in a very strange form. Cardano, for example, considered the equation $y^3 + 6y - 20 = 0$, for which the formula yields

$$y = \sqrt[3]{10 + \sqrt{108}} + \sqrt[3]{10 - \sqrt{108}}.$$

This form of the answer conceals the fact that the root is just $y = 2$. (It must be, since the equation has only one positive solution, and that solution is $y = 2$. You can verify with a calculator that this formula really does yield the number 2.)

Even worse things happen than what we have already described. Consider the equation $y^3 - 7y + 6 = 0$. The formula yields

$$y = \sqrt[3]{-3 + \sqrt{-100/27}} + \sqrt[3]{-3 - \sqrt{-100/27}}.$$

Thus the situation mentioned above has now arisen: We need to take the cube root of a complex number, that is, find a number $z = u + iv$ such that $z^3 = r + is$. In terms of real numbers, that means solving the equations $u^3 - 3v^2 u - r = 0$ and $v^3 - 3u^2 v + s = 0$. These equations are of the same form as our original equation, only now there are *two* of them. We can solve the second one for u in terms of v:

$$u = \sqrt{\frac{v^3 + s}{3v}}.$$

As you can see, this approach is hopeless. We don't yet know what v is going to be, but already it is appearing in a fraction under a square root sign, and we are going to have to cube u and insert it into the other equation. Let us surrender quickly and admit that our solution sometimes spins out of control. We conceal our ignorance by *fiat*, merely invoking the Cardano formula as if it solved the problem.

We conclude that what the cubic formula *really* says is the following: *If* we could extract the cube root of every complex number, *then* we could solve any cubic equation. But how do we extract the cube root of a complex number? In particular, how do we find a cube root of the complex number

$$-3 + \sqrt{-100/27} = -3 + \frac{10}{3\sqrt{3}}i?$$

3.1. Cube roots of complex numbers. Actually, we *can* extract the cube root of a complex number, but we can't do so using only real algebraic operations on real numbers. We have to resort to trigonometry. It turns out that every complex number $z = x + iy$ can be written in the *polar* form $z = r(\cos\theta + i\sin\theta)$, where $r = \sqrt{x^2 + y^2}$ is a positive number and

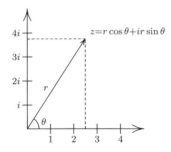

FIGURE 10. Polar representation of the complex number $z = x + iy = r\cos\theta + ir\sin\theta$.

$\theta = \arctan(y/x)$ is a certain angle, as shown in Fig. 10. One cube root of z is $z^{1/3} = \sqrt[3]{r}\left(\cos(\theta/3) + i\sin(\theta/3)\right)$. (There are two others, found by adding $120°$ and $240°$ to $\theta/3$.) For the complex number $-3 + \frac{10}{3\sqrt{3}}i$ mentioned above, we have $r = \sqrt{9 + 100/27} = \sqrt{343/27} = 7\sqrt{21}/9 \approx 3.56422554$ and $\theta = \arctan\left(-10/(9\sqrt{3})\right) \approx 2.5712158$ radians or 147.31981 degrees. Hence the cube root is approximately $\sqrt[3]{3.56422554}\left(\cos(0.857072) + i\sin(0.857072)\right)$, which computation reveals to be approximately $1 + 1.154701i$.

As just shown, extracting the cube root of a complex number involves taking the cube root of the positive real number r, which (as we have mentioned) is the ancient problem of *doubling the cube* when $r = 2$, and *trisecting the angle* θ, which is another ancient problem. The Greeks solved both of these problems using hyperbolas and parabolas. No wonder Umar al-Khayyam was successful in solving cubic equations by use of conic sections. It is remarkable that these two ancient geometric problems, studied long before algebra was invented, should turn out to be the key to solving a fundamental problem in algebra.

4. Alternative forms of the cubic formula

A further theoretical problem arises, since every complex number has *two* square roots (negatives of each other) and *three* cube roots, differing by factors of α and α^2, where $\alpha = -1/2 + (\sqrt{3}/2)i$ is a *primitive* cube root of 1, that is, its powers α, α^2, $\alpha^3 = 1$ are all the cube roots of 1. The formula appears to require that we choose one of these square roots and then one of these cube roots in two different places. That would seem to give us 36 possibly different values for the root. Actually, as the formula shows, both square roots give the same formula, since replacing the square root by its negative merely interchanges the two terms. That would still seem to leave nine possibilities for the root. However, the product of the two terms is a cube root of $-p^3/27$, and hence can assume only the three values $-p/3$, $-\alpha p/3$, and $-\alpha^2 p/3$. Substitution of this formula into the polynomial leads to an expression that is identically zero if and only if the two cube roots it contains are chosen so that their product is $-p/3$. In other words, the formula gives an actual root of the equation only if we use the "rules" $\sqrt[3]{a}\sqrt[3]{b} = \sqrt[3]{ab}$ (instead of the correct rule $\sqrt[3]{a}\sqrt[3]{b} = \alpha^j\sqrt[3]{ab}$ for some $j = 0, 1, 2$) and $\sqrt[3]{a^3} = a$ (instead of the correct rule $\sqrt[3]{a^3} = \alpha^j a$ for some $j = 0, 1, 2$). Thus, in fact the choice of one of the two cube roots determines the choice of the other, and so we have, as we should have, only three roots for this cubic polynomial.

To simplify the use of the cubic formula, we can exploit the fact that the product of the two cube roots must be $-p/3$ and write the formula as

$$\frac{p}{3\sqrt[3]{q/2 + \sqrt{q^2/4 + p^3/27}}} - \sqrt[3]{\frac{q}{2} + \sqrt{\frac{q^2}{4} + \frac{p^3}{27}}},$$

or, alternatively

$$\frac{p}{3(q/2 + \sqrt{q^2/4 + p^3/27})} z^2 - z,$$

where

$$z = \sqrt[3]{\frac{q}{2} + \sqrt{\frac{q^2}{4} + \frac{p^3}{27}}}.$$

This formula is the promised analog of the quadratic formula given in the previous chapter.

When complex numbers are substituted for a variable z in the formula $\sqrt{z}$, the formula becomes ambiguous, since every complex number except 0 has two square roots. Consequently, when we apply the cubic formula with specific numbers in place of p and q, we might expect to get two different roots by choosing the two possible values of the square root. But, as we have just seen, that doesn't happen. Choosing the opposite value for the square root merely interchanges the two terms in the sum, leaving the formula unchanged. On the other hand, each of the three possible values of the cube root in the formula will produce a different root of the polynomial. Thus, at last, we have arrived at a complete solution of the cubic equation, assuming that one can take the cube root of a complex number, as a formula in terms of the coefficients that becomes identically zero when substituted for the variable of the polynomial.

5. The "irreducible case"

For which cubic equations does the cubic formula involve the cube root of a complex number? It turns out that when the coefficients p and q are real, this happens precisely when there are three distinct real roots u, v, and w. The formula "gets confused," since it doesn't "know" which of the roots to choose. What can happen depends on the *discriminant* $(u-v)^2(v-w)^2(w-u)^2$. Since this discriminant is symmetric in the roots, we know that it can be computed from the coefficients. The resulting expression for the discriminant in terms of the coefficients is

$$\frac{-27a^2d^2 + 18abcd - 4ac^3 - 4b^3d + b^2c^2}{a^4}.$$

In the case when $a = 1$, $b = 0$, $c = p$, and $d = q$, this expression equals $-108(p^3/27+q^2/4)$. As we know, this "standardized" case is obtained by the substitution $y = x - b/3a$, which shifts all three roots by the same amount, namely $(u + v + w)/3$. The differences of the pairs of roots are unaffected by this change of variable, so that the discriminant is the same for both the equation in x and the equation in y.

As with quadratic polynomials, the term *discriminant* is used with a slight ambiguity. While the actual discriminant is $(u-v)^2(v-w)^2(w-u)^2$, we define its product by $-1/108$ as the *cubic discriminant* $D_3 = p^3/27 + q^2/4$. The expression for D_3 in terms of the roots shows that if the roots u, v, and w are real and distinct, the cubic discriminant will be negative, and hence the cubic formula will require the cube root of a complex number. On the other hand, if one of the roots, say u, is real and the other two are a pair of conjugate complex numbers, that is, $w = \bar{v}$, where $v = v_1 + iv_2$, then $D_3 = (1/27)\left((u - v_1)^2 + v_2^2\right)^2 v_2^2$, which is positive. Thus we see that for a cubic polynomial with real coefficients, the Cardano formula requires complex numbers *if and only if* all three roots are real and distinct.

As the expression for the discriminant in terms of the roots shows, when the discriminant is zero, the equation $y^3 + py + q = 0$ has either three equal real roots (a triple root) or a single root and a double root. In that case, if p and q are real, the cubic formula with real cube roots taken yields the triple root (which must be zero, since the sum of the roots is zero) or the single root, *never* the double root. Thus our preference for real cube roots of real numbers allows us to break the symmetry of the cubic formula in some cases and distinguish one root from another.

5.1. Imaginary numbers. To repeat what we have just said for emphasis, *only when the roots are all three real and distinct does the cubic discriminant become negative for real p and q.* And therein lies the reason for introducing imaginary and complex numbers into mathematics. While it might be all right to say that a quadratic equation has no roots if the quadratic formula requires the square root of a negative number, that answer will not do for the equation $y^3 - 7y + 6 = 0$, whose cubic discriminant is negative, but which has real roots $y = 1$, $y = 2$, $y = -3$. As a result of this cubic formula, mathematicians began to make sense of imaginary and complex numbers. Rafael Bombelli, mentioned above, was able to show that the cube roots of complex numbers in the cubic formula could be chosen so that the formula really would yield the three real-valued solutions of the equation.

Bombelli's work was not a systematic introduction of complex numbers as a subject of study. He used them only as a way of getting real numbers in the end. The case of three distinct real roots came to be known as the *irreducible* case, and mathematicians like François Viète (1540–1603) tried to find methods of solving it involving only real numbers. Viète managed to do so, again using trigonometry (trisecting an angle). But trisecting an angle requires solving another irreducible equation; and in any case, the introduction of trigonometry into the solution takes the solution outside the realm of pure algebra. Later mathematicians were able to show that there is no *algebraic* formula for the solution of a general cubic equation that involves only real numbers for every equation with three real roots. The path from coefficients to roots begins and ends in the real numbers, but cannot be confined to them; it must take a detour through the complex numbers. In this way, the cubic equation led to the acceptance of complex numbers, which have proved to be of immense value in the most diverse areas of mathematics and physics.

6. Problems and questions

Problem 7.1. Change the equation $y^3 + py + q = 0$ to an equation of the form $w^3 + w^2 = N$ by the substitution $y = q/(pw)$.

Problem 7.2. List in "generic" form all the possible cubic equations, given that coefficients must be positive numbers, for example, "cubes plus first-degree terms equal constants." Note that there must be both cubes and constant terms. Otherwise the equation is not really a cubic equation.

Problem 7.3. Given an equation $Ax^3 + Bx^2 + Cx = D$, where A, B, C, and D are positive numbers, show how to choose positive numbers a, b, and c so that this equation is the same as $x^3 + ax^2 + b^2x = b^2c$. Then show that its only positive solution is one of the x coordinates of the points of intersection of the hyperbola and circle

$$xy = bc,$$
$$\left(x + \frac{a-c}{2}\right)^2 + (y-b)^2 = \left(\frac{a+c}{2}\right)^2.$$

Problem 7.4. Use the technique of the previous problem to reduce the solution of the equation $x^3 + 3x^2 + 15x = 27$ to finding the intersection points of a circle and a hyperbola. Can you estimate the solution graphically?

Problem 7.5. Solve the following equations using the Cardano formula for a cubic (use a calculator if you need the cube root of a complex number):

1. $x^3 + 153x - 4886 = 0$,
2. $x^3 - 6x^2 + 144x - 1539 = 0$ (you must first eliminate the squared term),
3. $x^3 - x - 1 = 0$,
4. $x^3 - 6x^2 + 11x - 6 = 0$.

Problem 7.6. Show that if α is a primitive pth root of unity, where p is a prime, so is α^j for any j that is not a multiple of p.

Problem 7.7. List all the fourth roots of unity. Which of them are primitive?

Question 7.1. We saw in the previous lesson that the algebraic relation $z = \sqrt{w}$ between two complex variables $z = u + iv$ and $w = r + is$ can be written as

$$r = u^2 - v^2; \quad s = 2uv,$$

and that this relation can be inverted to yield

$$u = \sqrt{\frac{\sqrt{r^2 + s^2} + r}{2}}; \quad v = \sqrt{\frac{\sqrt{r^2 + s^2} - r}{2}}.$$

Similarly, the algebraic relation $z = \sqrt[3]{w}$ can be written as a pair of algebraic relations,

$$r = u^3 - 3uv^2; \quad s = 3u^2v - v^3.$$

Again, these are relations among *real* variables. What is the essential difference between these two cases that makes the cubic equation noticeably more difficult than the quadratic?

Question 7.2. What is the difference between cubic equations with rational solutions, like $x^3 + 45x - 98 = 0$, for which the Cardano formula yields a recognizable, familiar number, and equations like $x^3 + 17x - 42 = 0$, for which the same answer appears in a strange form? How can you tell which of these is likely to happen?

7. Further reading

Daoud S. Kasir, *The Algebra of Omar Khayyam*, Columbia University Press, New York, 1931.

R. B. McClenon, "Cardan. Solution of the cubic equation," in *Source Book in Mathematics*, David Eugene Smith, ed., Dover, New York, 1959.

Dirk J. Struik, "Cardan. On cubic equations," in *Source Book in Mathematics, 1200–1800*," D. J. Struik, ed., Princeton University Press, Princeton, NJ, 1986.

Part 3

Resolvents

When the combinatorial approach was applied to equations of higher degree than the third, it was found that an appropriate substitution to simplify the quartic equation could not be found without solving a certain cubic equation. The combinatorial technique reached its high-water mark with the solution of this problem. It could do no more, but the effort to make it work produced a general strategy for finding roots via a *resolvent*, a function of the roots of an equation that assumes fewer values than there are roots when the roots are permuted. For a cubic equation with roots u, v, w, the function $(u + \alpha v + \alpha^2 w)^3$ (where $\alpha = -1/2 + (\sqrt{3}/2)i$ is a complex cube root of unity) assumes only two different values when u, v, w are permuted, since a cyclic permutation of the three roots is tantamount to multiplying the function by α^3 or α^6, both of which equal 1. A resolvent satisfies an equation of lower degree than the original equation, and the coefficients of that equation are symmetric functions of the roots of the original equation. They can be expressed in terms of the coefficients of the original equation, which are known. In this way, the resolvent provides additional information about the roots, information that can be obtained by solving an equation of lower degree. As its name implies, it helps to solve the equation. Lessons 8 and 9 explore how resolvents arose, examine several different resolvents for a quartic equation, and consider the challenge of finding a resolvent for the quintic equation.

From Combinatorics to Resolvents

The formulaic solution of the cubic equation was a major milestone in the history of algebra. By analyzing this solution, mathematicians came to understand what was involved in expressing the roots of a polynomial in terms of its coefficients. Eventually, this understanding led to a proof that no *algebraic* formula could be given for solving the quintic equation. The concepts of group and field that were engendered by this research have proved to be even more valuable than the problem that gave rise to them. All this was beyond the horizon in the midsixteenth century, when the first formula was given. That formula led to immediate progress. As we have already noted, it led to the acceptance of complex numbers, another valuable analytic tool of modern science.

It was obvious that matters were going to become more complicated as attempts were made to solve equations of ever higher degrees. Up to this point, the approach had been combinatoric: algebraic substitutions were sought that would reduce the equations to a simpler form. But these substitutions become progressively harder to find. The quadratic equation $ax^2 + bx + c = 0$ is reduced to the extraction of a square root by the substitution $z = x - b/2a$. For the cubic $ax^3 + bx^2 + cx + d = 0$, the corresponding substitution $z = x - b/3a$ removes one term, but must then be combined with the identity $(u - v)^3 + 3uv(u - v) + (v^3 - u^3) \equiv 0$. That identity produces u and v via a quadratic equation for u^3 or v^3 and allows z to be expressed as $u - v$. The highest-degree equation for which such a combinatorial technique will work is the quartic, and for that case the appropriate substitution is found by solving a cubic equation.

The systematic search for a substitution that would eliminate both the linear and quadratic terms in the cubic, and the attempt to perform a similar rearrangement to reduce the quartic equation to root extractions led mathematicians to the concept of a *resolvent*, an asymmetric function of the roots that assumes fewer values when the roots are permuted than there are roots. The resolvent satisfies an equation of lower degree than the original, but the coefficients of that equation are symmetric in the roots of the original and hence can be expressed in terms of those of the equation being studied. In this way, a general strategy for solving equations arises: start with the symmetric functions of the roots represented by the coefficients, for example, $f(s, t, u, v) = s + t + u + v$ in the case of the quartic, and asymmetrize them via resolvents in order to produce the highly nonsymmetric function $F(s, t, u, v) = s$, which is a root.

Even before the Cardano solution of the cubic was fully sorted out, Cardano and his student Ferrari had cracked the problem of the quartic equation, showing how the general quartic could be solved by a substitution that could be discovered by solving a particular cubic equation. This procedure turned out to be a much smaller step than the solution of the cubic had been. At the same time, as mentioned in the preceding lesson, François Viète found a trigonometric solution of the irreducible case of the cubic that avoided the use of complex numbers. Ehrenfried Walther von Tschirnhaus (1652–1708) discovered that, just as the quadratic equation $ax^2 + bx + c$ can be reduced to a pure equation $z^2 = N$ by the substitution $z = x + b/(2a)$, a general cubic $y^3 + py + q = 0$ can be reduced to a pure equation $z^3 = N$ by a substitution of the form $z = y^2 + ry + s$ if r and s are suitably chosen (see Problem 8.5), and that the suitable choice can be found by solving only linear and quadratic equations. Thus, the quadratic can be solved by a substitution that can be found by solving a linear equation; the cubic, by a substitution that can be found by solving a quadratic equation; and the quartic, as Cardano and Ferrari had shown, by a substitution that can be found by solving a cubic equation. These facts suggested to Tschirnhaus that a "bootstrapping" technique might be possible, enabling the solution of equations of degree n to be reduced to extracting nth roots and using substitutions that can be found by solving equations of degree $n - 1$. Indeed, he said as much in a letter to Gottfried Wilhelm von Leibniz (1646–1716) in 1677. But he was mistaken, as we shall see below (Problem 8.7).

1. Solution of the irreducible case using complex numbers

In the preceding lesson, when we were trying to solve the equation $y^3 - 7y + 6 = 0$ using the cubic formula, we found that we needed the cube root of the complex number $-3 + \frac{10}{3\sqrt{3}}i$. How do we find this cube root? We gave a geometric method in the preceding lesson and found an approximate value $1 + 1.154701i$. But in this case we can pull the "exact" value out of our hat by inspired guessing: $z = 1 + \frac{2}{\sqrt{3}}i$ will do, as you can verify by direct computation. Rafael Bombelli showed that if p and q are real numbers and $z = r + si$ is a complex number such that

$$z^3 = (r^3 - 3s^2 r) + (3r^2 s - s^3)i = -\frac{q}{2} + \sqrt{\frac{p^3}{27} + \frac{q^2}{4}},$$

then $z\alpha$ and $z\alpha^2$ are also cube roots of this number, where $\alpha = -1/2 + (\sqrt{3}/2)i$ and $\alpha^2 = -1/2 - (\sqrt{3}/2)i$ are the two nonreal cube roots of 1. Then the three roots of the equation $y^3 + py + q = 0$ are $(r + si) + (r - si) = 2r$, $(r + si)\alpha + (r - si)\alpha^2 = -r - \sqrt{3}s$, and $(r + si)\alpha^2 + (r - si)\alpha = -r + \sqrt{3}s$. In the present case, $r = 1$ and $s = 2/\sqrt{3}$, so that the roots are 2, −3 and 1. But

Bombelli introduced complex numbers minimally, as a formal way of making sense of the cubic formula. The imaginary parts had to drop out at the end, leaving the three real solutions. Complex numbers themselves were not at first recognized as solutions, although nowadays we learn to accept them from our first algebra course as the solutions of certain quadratic equations. Obviously, they were also not considered as coefficients in a polynomial.

2. The quartic equation

The general quartic equation is $ax^4 + bx^3 + cx^2 + dx + e = 0$, and as usual, division by a and the substitution $x = y - b/(4a)$ reduces this equation to the form

$$y^4 = py^2 + qy + r.$$

We transferred the last three terms to the right side here for convenience. Now the idea is to add $2ty^2 + t^2$ to both sides. The left side will then be $(y^2 + t)^2$. If t is suitably chosen, the right side will also be a perfect square. The question is, how do we choose t?

Thus we have

$$(y^2 + t)^2 = y^4 + 2ty^2 + t^2 = (p + 2t)y^2 + qy + (r + t^2).$$

But, as discussed in Lesson 6, a quadratic polynomial $ay^2 + by + c$ is a perfect square if and only if its quadratic discriminant $D_2 = b^2 - 4ac$ is zero. Thus t must be chosen so that

$$q^2 - 4(p + 2t)(r + t^2) = 0.$$

As you see, this is a *cubic* equation in t, called the *resolvent* cubic: namely $8t^3 + 4pt^2 + 8rt + 4pr - q^2 = 0$. If t is chosen so as to satisfy that equation, then the equation we are trying to solve can be rewritten as

$$(y^2 + t)^2 = (p + 2t)\left(y + \frac{q}{2(p + 2t)}\right)^2,$$

and our original quartic equation breaks into the two quadratic equations

$$y^2 + t = \pm\sqrt{p + 2t}\left(y + \frac{q}{2(p + 2t)}\right).$$

Why does this technique work? How does it happen that the substitution we need can be found by solving a cubic equation? Careful analysis of the procedure, involving some very messy computation, reveals that if the four roots of the original quartic equation are w, x, y, and z, then one root of the resolvent equation is

$$t_1 = \frac{-(w + x + y + z)^2 + 8(wx + yz)}{16}.$$

The other two roots t_2 and t_3 can be obtained by interchanging x with y and x with z. The expression on the right side of the last equation can assume only these three values when the four roots are permuted among themselves. It follows that the cubic polynomial $(t - t_1)(t - t_2)(t - t_3)$ has coefficients that are symmetric in w, x, y, and z, and hence expressible in terms of p,

q, and r. For that reason, we shall transfer the name *resolvent* from this auxiliary equation and apply it to any nonsymmetric rational function of the roots that assumes fewer values than there are roots when the roots are permuted. (The symmetric functions are the coefficients we started with. They do have some claim to be called resolvents, since they give information about the roots. But they are given at the outset, and so we shall reserve the term for nonsymmetric functions.)

To fix these ideas, let us consider as an example the equation $x^4 - 6x^2 - 8x + 24 = 0$, so that $y = x$. We write this equation as $x^4 = 6x^2 + 8x - 24$. We want to choose t so that the equation will assume the form

$$(x^2 + t)^2 = (6 + 2t)\left(x + \frac{4}{6 + 2t}\right)^2.$$

The resolvent cubic is $8t^3 + 24t^2 - 192t - 640 = 0$, which is equivalent to $t^3 + 3t^2 - 24t - 80 = 0$. The substitution $y = t+1$, that is, $t = y-1$, reduces this equation to $y^3 - 27y - 54 = 0$. This happens to be an equation for which the Cardano formula works well. The formula discloses that $y = 6$, and therefore $t = 5$. Hence our original equation breaks into two quadratic equations

$$x^2 + 5 = \pm 4\left(x + \frac{1}{4}\right) = \pm(4x + 1),$$

that is, $x^2 - 4x + 4 = 0$ or $x^2 + 4x + 6 = 0$. Therefore $x = 2$ (a double root) or $x = -2 \pm \sqrt{2}i$.

In general, though, if we needed numerical answers, it would be more efficient to use numerical methods like those of the Chinese. This formula is far too cumbersome to be practical. Its chief value is that it formed part of the search for a general way of solving all polynomial equations.

3. Viète's solution of the irreducible case of the cubic

Although complex numbers began to gain acceptance after the work of Cardano and Bombelli, attempts were still made to solve the irreducible case using only real numbers. As already mentioned, this case is easily identified because the cubic discriminant $D_3 = p^3/27 + q^2/4$ is negative. It turns out that trigonometry provides the answer to this problem, as François Viète discovered.

The classical problem of trisecting the angle reduces to a cubic equation through the trigonometric identity

$$\cos^3\left(\frac{\theta}{3}\right) - \frac{3}{4}\cos\left(\frac{\theta}{3}\right) - \frac{1}{4}\cos\theta = 0.$$

This is a cubic equation in the variable $y = \cos(\theta/3)$. Its cubic discriminant is $D_3 = (-1 + \cos^2\theta)/64$, that is, $-\sin^2\theta/64$, which is negative for all nontrivial angles θ. (Only for multiples of π does it become zero, indicating that two of the roots are equal.)

To solve the irreducible equation $y^3 + py + q = 0$, Viète's idea was to scale y by setting $z = (\sqrt{3}/(2\sqrt{-p}))y$. (The negative sign is necessary since $p < 0$ in the irreducible case.) We then get the equation

$$z^3 - \frac{3}{4}z - \frac{1}{4}\frac{3\sqrt{3}q}{2p\sqrt{-p}} = 0.$$

As a result, we could choose $z = \cos(\theta/3)$ if we could find an angle θ such that

$$\cos\theta = \frac{3\sqrt{3}q}{2p\sqrt{-p}}.$$

This will be possible provided $27q^2/(-4p^3) \leq 1$. In other words $q^2/4 \leq -p^3/27$, which just happens to be the condition that the cubic discriminant be nonpositive!

As an example, let us return to the equation $y^3 - 7y + 6 = 0$. The corresponding equation for z is

$$z^3 - \frac{3}{4}z + \frac{1}{4}\frac{9\sqrt{3}}{7\sqrt{7}} = 0.$$

We need $\cos\theta = -9\sqrt{3}/(7\sqrt{7})$. This gives $\theta \approx 147.32° \approx 2.57$ radians, and so the three possible values of $\theta/3$ are $49.1066° \approx 0.86$ radians, $169.1066° \approx 2.95$ radians, and $-70.8934° \approx -1.23$ radians. The cosines of these angles yield the three values of z: 0.654654, -0.981981, and 0.327327. Finally, the corresponding values of $y = z\sqrt{28/3}$ are 2, -3, and 1.

3.1. Comparison of the Viète and Cardano solutions.

Viète's solution uses trigonometry to find the cube root of a complex number. When p and q are real, the imaginary parts of the two complex cube roots cancel each other, and so we get a purely real expression for the solution. Viète's method amounts to choosing the angle θ so that

$$\cos\theta = \frac{3\sqrt{3}}{-p\sqrt{-p}}\left(-\frac{q}{2}\right),$$

so that

$$-\frac{q}{2} = \frac{-p\sqrt{-p}}{3\sqrt{3}}\cos\theta = \sqrt{\frac{-p^3}{27}}\cos\theta.$$

It then follows that

$$\sin\theta = \sqrt{1 - \cos^2\theta} = \sqrt{1 + \frac{27q^2}{4p^3}} = \sqrt{\frac{27}{p^3}}\sqrt{\frac{q^2}{4} + \frac{p^3}{27}},$$

that is,

$$\sqrt{\frac{q^2}{4} + \frac{p^3}{27}} = i\sqrt{\frac{-p^3}{27}}\sin\theta.$$

Hence

$$
\begin{aligned}
y &= \sqrt[3]{-\frac{q}{2} + \sqrt{\frac{q^2}{4} + \frac{p^3}{27}}} + \sqrt[3]{-\frac{q}{2} - \sqrt{\frac{q^2}{4} + \frac{p^3}{27}}} \\[2mm]
&= \sqrt[3]{\sqrt{\frac{-p^3}{27}}\,(\cos\theta + i\sin\theta)} + \sqrt[3]{\sqrt{\frac{-p^3}{27}}\,(\cos\theta - i\sin\theta)} \\[2mm]
&= \sqrt{\frac{-p}{3}}\left(\cos\left(\frac{\theta}{3}\right) + i\sin\left(\frac{\theta}{3}\right) + \cos\left(\frac{\theta}{3}\right) - i\sin\left(\frac{\theta}{3}\right)\right) \\[2mm]
&= \frac{2\sqrt{-p}}{\sqrt{3}}\cos\left(\frac{\theta}{3}\right).
\end{aligned}
$$

Thus, Viète's solution represents a condensation of the Cardano solution. It represents the solution of the equation $y^3 + py + q = 0$ in *transcendental form*:

$$
y = \frac{2\sqrt{-p}}{\sqrt{3}}\cos\left(\frac{1}{3}\arccos\left(\frac{3\sqrt{3}q}{2p\sqrt{-p}}\right)\right).
$$

In that form, it applies to all complex values of p and q and shows that *an algebraic relation can be equivalent to a transcendental relation*. This phenomenon was encountered again in the eighteenth century, as mathematicians worked out the theory of elliptic integrals, which are transcendental functions that nevertheless have algebraic addition formulas. (One example of an elliptic function is the Jacobi amplitude function mentioned in Lesson 4.) As it eventually turned out, the simplest solution of a general quintic equation can be expressed in terms of elliptic functions.

A final remark: The inverse cosine function "arccos" is multivalued. It has only one value (its principal value) between 0 and π, but any multiple of 2π may be added to it.

4. The Tschirnhaus solution of the cubic equation

Tschirnhaus found a second approach to the cubic equation, an approach he hoped to generalize to equations of all degrees. His idea was to find some new variable z in terms of which the equation $y^3 + py + q = 0$ would have the form $z^3 = N$, and hence the obvious solution $z = \sqrt[3]{N}$. And he succeeded. We shall list the steps in his method as an algorithm.

1. Let $z = y^2 + ry + s$, where r and s are chosen by solving the system

$$
\begin{aligned}
2p - 3s &= 0, \\[2mm]
pr^2 + 3qr - \frac{p^2}{3} &= 0,
\end{aligned}
$$

that is,

$$r = \frac{3}{p}\left(-\frac{q}{2} + \sqrt{\frac{p^3}{27} + \frac{q^2}{4}}\right),$$

$$s = \frac{2p}{3}.$$

Tschirnhaus chose these values after rewriting the equation for y in terms of z and setting the coefficients of z^2 and z equal to zero. (For more details, work Problem 8.5 below.) Those coefficients are $2p - 3s$ and $pr^2 + 3qr + p^2 - 4ps + 3s^2$, which becomes $pr^2 + 3qr - p^2/3$ when $s = 2p/3$. Notice that r is just $3/p$ times the cube of one of the two terms in the Cardano formula. In particular, r contains the square root of the cubic discriminant, and hence is a *complex* number if the equation has three distinct real roots. Thus, the solution given by Tschirnhaus also involves a complex number in this case.

2. When y is eliminated between the two equations, the result is an expression that contains z under a radical (as discussed below). When that radical is removed by symmetrizing (multiplying by the conjugate radical) and these values are used for r and s, the equation satisfied by z is the *pure* cubic equation

$$z^3 = \frac{2p^3}{27} + q^2 + pqr + \frac{2p^2r^2}{3} - qr^3.$$

3. Find z by extracting the cube root of both sides in this last equation.

4. Solve the quadratic equation $y^2 + ry + (s - z) = 0$ to find y.

As you can see, the Tschirnhaus method lacks the symmetry of the Cardano method and is therefore much harder to remember. It also has a second disadvantage, as we are about to see. When applying it, one can skip the explanations given above and just compute sequentially r, s, z, and y from the formulas given.

When the parameters have specific values, we might expect some simplification. Let us consider a case where the Cardano formula works well, the equation $y^3 + 105y - 218 = 0$. Thus we have $s = 70$, and r satisfies $105r^2 - 654r - 3675 = 0$. With a little trouble, we find that $r = \frac{49}{5}$ or $r = -\frac{25}{7}$. Let us keep things simple and take the first of these as the value of r. We next find

$$z = \sqrt[3]{\frac{2p^3}{27} + q^2 + pqr + \frac{2p^2r^2}{3} - qr^3} = \frac{468}{5}.$$

We can then find y by solving the equation

$$y^2 + (49/5)y - (118/5) = 0,$$

getting $y = 2$ or $y = -\frac{59}{5}$. The first of these satisfies the original equation, the second does not. If we had chosen $r = -\frac{25}{7}$, instead, we would have found $z = \frac{468}{7}$. The equation for y is then

$$y^2 - (25/7)y + (22/7) = 0,$$

whose roots are $y = 2$, $y = \frac{11}{7}$. Thus, either of the possible values of r leads to the correct solution of the equation, but also to an extraneous root. And therein lies the second disadvantage of the Tschirnhaus method.

These extraneous roots enter because the original expression for y, namely

$$y = \frac{-r + \sqrt{r^2 - 4s + 4z}}{2},$$

involves the variable z under a radical sign. The new equation is of the form $(Az + B) + (Cz + D)\sqrt{Ez + F} = 0$. In order to get a polynomial equation in z, it is necessary to multiply by the conjugate expression $(Az + B) - (Cz + D)\sqrt{Ez + F}$, and that is where the extraneous roots enter. For another derivation of this equation, one that does not use radicals, see Problem 8.5.)

5. Lagrange's reflections on the cubic equation

We now continue the more abstract study of general polynomial equations one step beyond what we did for quadratic equations in Lesson 6. Consistent with the notation we used there, we write a typical cubic polynomial as $x^3 - ax^2 + bx - c$. We know that if the roots are u, v, and w, then $a = u+v+w$, $b = uv+vw+uw$, and $c = uvw$. Just as in the case of the quadratic equation, any formula that contains the roots symmetrically can be written in terms of a, b, and c. For example, $u^2 + v^2 + w^2 = a^2 - 2b$, $u^3 + v^3 + w^3 = a^3 - 3ab + 3c$, $u^2v + u^2w + uv^2 + uw^2 + v^2w + vw^2 = ab - 3c$, and so on.

As we saw, the secret of finding the roots is to make the substitution $x = y + a/3$, leading to the "standardized" cubic polynomial $y^3 + py + q$, with $p = b - a^2/3$ and $q = -2a^3/27 + ab/3 - c$. For this polynomial the formula

$$x = \frac{a}{3} + z + \frac{p}{3(q/2 + \sqrt{q^2/4 + p^3/27})} z^2,$$

where

$$z = \sqrt[3]{-\frac{q}{2} - \sqrt{\frac{q^2}{4} + \frac{p^3}{27}}},$$

when substituted for x in the polynomial $x^3 - ax^2 + bx - c$, yields the function

$$\left(-\frac{p^2}{3\left(q/2 + \sqrt{q^2/4 + p^3/27}\right)} + \frac{p^2}{3\left(q/2 + \sqrt{q^2/4 + p^3/27}\right)}\right) z^2 + (p - p)z$$
$$+ \left(\frac{p^3 - 27\left(q/2 + \sqrt{q^2/4 + p^3/27}\right)^2 + 27q\left(q/2 + \sqrt{q^2/4 + p^3/27}\right)}{27\left(q/2 + \sqrt{q^2/4 + p^3/27}\right)}\right).$$

This formula is obtained using only the relation $z^3 = -q/2 - \sqrt{q^2/4 + p^3/27}$. Even though there are normally three values of z that satisfy this relation when p and q are given specific numerical values, all three work equally well. Just as in the case of the quadratic equation, the coefficients of z^2 and z and the term independent of z are all identically, formally equal to 0 as expressions in p and q, and hence also as expressions in a, b, and c. It follows that when numerical values are assigned to a, b, and c, each of the three possible values of z will yield this same function of z, and so the result is always a root of the polynomial $x^3 - ax^2 + bx - c$.

5.1. The cubic formula in terms of the roots. The significant fact that the radicals in the quadratic formula become rational functions (without root extractions) when expressed in terms of the roots of the equation turns out to be true also in the case of the cubic formula. That is, the variable we called z above is a polynomial in the roots u, v, and w of the original polynomial.

To see what this polynomial is, note that there are two possible choices for the square root inside the cube root:

$$\sqrt{\frac{p^3}{27} + \frac{q^2}{4}} = \pm \frac{(u-v)(v-w)(w-u)}{6\sqrt{3}} i.$$

This relation shows that the square root inside the cube root is a polynomial in the roots of the original polynomial.

Next, although the computation is tedious, one can compute that if $\alpha = -1/2 + (\sqrt{3}/2)i$ is a primitive cube root of 1, then

$$z^3 = \left(\frac{u + \alpha v + \alpha^2 w}{3}\right)^3 = \left(\frac{w + \alpha u + \alpha^2 v}{3}\right)^3 = \left(\frac{v + \alpha w + \alpha^2 u}{3}\right)^3$$

if the positive sign is chosen for the square root and

$$z^3 = \left(\frac{u + \alpha^2 v + \alpha w}{3}\right)^3 = \left(\frac{w + \alpha^2 u + \alpha v}{3}\right)^3 = \left(\frac{v + \alpha^2 w + \alpha u}{3}\right)^3$$

if the negative sign is chosen. In other words, the six possible values of the expression

$$z = \sqrt[3]{-\frac{q}{2} - \sqrt{\frac{p^3}{27} + \frac{q^2}{4}}}$$

are precisely the six values obtained by permuting u, v, and w in the expression

$$\frac{u + \alpha v + \alpha^2 w}{3}.$$

Notice that this last formula is the exact analog of what we discovered in the case of the quadratic equation. This fact is a discovery of tremendous importance in the history of algebra: *The radicals in the formulas for solving quadratic and cubic equations in terms of the coefficients become polynomials when expressed in terms of the solutions of the equations, and the different values of the roots that must be extracted correspond to permutations of the set of solutions.*

The polynomials we have encountered here are not just *any* polynomials. They are, first, the discriminant, which tells us when two solutions are equal, and second, a linear combination of the solutions whose coefficients are the roots of unity, divided by the degree of the equation.

We have seen that the same is true of the quadratic equation. The function $(\alpha_1 u + \alpha_2 v)/2$ in the case of the quadratic polynomial, where $\alpha_1 = 1$ and $\alpha_2 = -1$ are the two square roots of unity, and the function $E = (\alpha_1 u + \alpha_2 v + \alpha_3 w)/3$ in the case of the cubic, where $\alpha_1 = 1$, $\alpha_2 = -1/2 + (\sqrt{3}/2)i$ and $\alpha_3 = -1/2 - (\sqrt{3}/2)i$ are the three cube roots of unity, are *not* symmetric functions of the roots. Nevertheless, as we have seen, they can be computed from the coefficients of the polynomial by first generating the discriminant using arithmetical operations, then taking its square root. In the case of the cubic, it is then necessary to add $-q/2$ and take a cube root. These nonsymmetric functions of the roots can be computed from the coefficients of the equation using root extractions. They provide a second *nonsymmetric linear* equation for the three roots, supplementing the symmetric equation that comes from the coefficient of x (in the case of the quadratic) or x^2 (in the case of the cubic). Thus we have:

$$u + v + w = a,$$
$$\alpha_1 u + \alpha_2 v + \alpha_3 w = A = 3E.$$

All we need is one more such equation, say

$$\alpha_1^2 u + \alpha_2^2 v + \alpha_3^2 w = B,$$

and we would have a complete system of *linear* equations for u, v, and w, in fact a system of the type mentioned in Lesson 2, in which the coefficients form a Vandermonde matrix. The problem is to write B in terms of the coefficients. That is easy to do, since $\alpha_3^2 = \alpha_2$ and $\alpha_2^2 = \alpha_3$, while $\alpha_1^2 = \alpha_1$. Thus the third equation that we desire can be obtained from the expression for A by merely interchanging v and w, and this operation amounts to reversing the sign of the square root inside the cube root, as we just saw. In any case, this third equation is provided by solving the same quadratic equation that yielded the second equation here.

In both quadratic and cubic equations, an asymmetric linear combination of the roots that can assume $n!$ values when the roots are permuted has now been created by nesting root extractions starting with functions of the roots that are symmetric and hence expressible in terms of the coefficients. One is naturally led to conjecture that it is always possible to produce a system of linear equations equivalent to a given polynomial equation in this way.

5.2. A test case: The quartic. Let us test this conjecture in the case of the quartic equation.

As we already know, the general quartic equation

$$ax^4 + bx^3 + cx^2 + dx + e = 0$$

reduces via the substitution $y = x - b/(4a)$ and division by a to an equation of the form

$$y^4 = py^2 + qy + r,$$

where p, q, and r are polynomials in the variables b/a, c/a, d/a, and e/a. Thus no root extractions are needed to get this far.

This equation then reduces to either of two quadratic equations, one of which is

$$y^2 - \sqrt{p + 2t}\,y - \left(\frac{q}{2\sqrt{p + 2t}} - t\right) = 0,$$

provided t is chosen as a solution of the cubic equation

$$8t^3 + 4pt^2 + 8rt + 4pr - q^2 = 0.$$

Looking at the substitutions one at a time, we get the following sequence of algebraic operations to find the roots:

$$x = y + \frac{b}{a},$$

$$y = \frac{1}{2}\left(\sqrt{p + 2t} + \sqrt{p - 2t + \frac{2q}{\sqrt{p + 2t}}}\right),$$

$$t = \sqrt[3]{-\frac{Q}{2} + \sqrt{\frac{Q^2}{4} + \frac{P^3}{27}}} + \sqrt[3]{-\frac{Q}{2} - \sqrt{\frac{Q^2}{4} + \frac{P^3}{27}}}.$$

Here P and Q are polynomials in p, q, and r, and hence also polynomials in b/a, c/a, d/a, and e/a. Thus, we see that the expression for y contains a nested pair of square roots, inside which t occurs, and the expression for t contains a square root inside a cube root, so that the total order of root extractions, down to the bottom layer is $4 \cdot 3 \cdot 2$, in other words, $4!$, *exactly as we conjectured would be the case.*

All this was realized by Lagrange, who wrote a long essay on the current state of polynomial equations in 1770. Although that date is about a century after the main subject of the present lesson, which is the methods of Viète and Tschirnhaus, we have glanced into the future here in order to put all the different ways of solving cubic equations in proximity to one another.

6. Problems and questions

Problem 8.1. Solve the equation $y^3 - 39y + 70 = 0$ using the Cardano formula and computing the real part of the cube root of the complex number $-35 + 18\sqrt{3}i$.

Problem 8.2. Find an "exact" cube root of $-35 + 18\sqrt{3}i$ as a complex number of the form $r + s\sqrt{3}i$, where r and s are rational numbers. *Hint:* Look at a numerical approximation to this cube root first. That should tell you what r is, and then the equation for s becomes very simple.

Problem 8.3. Solve the quartic equation

$$x^4 - 6x^3 + 14x^2 - 20x + 8 = 0.$$

Find all four roots.

Problem 8.4. Use Viète's method to find numerical approximations to the three real roots of the equation $y^3 - (11 + \sqrt{30})y + (5\sqrt{6} + 6\sqrt{5}) = 0$. What do you *suspect* are the "exact" values of these three roots?

Problem 8.5. Consider the pair of equations

$$\begin{aligned} 0 &= y^3 + py + q, \\ z &= y^2 + ry + s. \end{aligned}$$

Multiplying the second equation by y and subtracting the first, derive the equation

$$zy = ry^2 + (s - p)y - q = r(z - ry - s) + (s - p)y - q,$$

and conclude that

$$y = \frac{rz - q - rs}{z + p + r^2 - s}.$$

This equation expresses y as a *fractional-linear* or *Möbius* transformation of z, so called after August Ferdinand Möbius (1790–1868). It is easy to verify that this equation can be solved to express z as a Möbius transformation of y, so that y and z are in one-to-one correspondence when the two equations written above are satisfied. An exception to this claim occurs when $r^3 + pr + q = 0$; in that case, every value of z yields the same value r for y, since the numerator of the fraction equals r times the denominator in that case. However—see Problem 8.11—the equation itself is trivial in that case.)

When this expression for y is substituted into the equation $y^3 + py + q = 0$ and the denominator is cleared, the result is the equation

$$(r^3 + pr + q)(z^3 + (2p - 3s)z^2 + (p^2 + 3qr + pr^2 - 4ps + 3s^2)z$$
$$+ (-q^2 + pqr + qr^3 - p^2s - 3qrs - pr^2s + 2ps^2 - s^3)) = 0.$$

The three values of z that make the second factor in this last equation equal to zero must correspond to the three values of y that satisfy the original equation. Thus, we now need to solve the cubic equation

$$z^3 + (2p - 3s)z^2 + (p^2 + 3qr + pr^2 - 4ps + 3s^2)z$$
$$+ (-q^2 + pqr + qr^3 - p^2s - 3qrs - pr^2s + 2ps^2 - s^3) = 0.$$

Obviously, choosing $s = 2p/3$ will cause the coefficient of z^2 to vanish. Then, solving the *quadratic* equation

$$pr^2 + 3qr - \frac{p^2}{3} = 0$$

for r and inserting this value of r will cause the coefficient of z to vanish, leaving a "pure" equation

$$z^3 - N = 0.$$

Once $z = \sqrt[3]{N}$ is found, we find y by solving the original quadratic equation

$$y^2 + ry + (s - z) = 0.$$

Work through this procedure step by step for the equation $y^3 + 18y + 30 = 0$. Observe that there are two possible values for r, namely $r = 1$ and $r = -6$, and that when you solve the first one for y with $r = 1$, you get the two values $y = \sqrt[3]{6} - \sqrt[3]{36} \approx -1.48481$ and $y = -1 - \sqrt[3]{6} + \sqrt[3]{36} \approx 0.48481$. When you solve for y with $r = -6$, you get the two values $y = \sqrt[3]{6} - \sqrt[3]{36} \approx -1.48481$ and $y = 6 - \sqrt[3]{6} + \sqrt[3]{36} \approx 7.48481$. How do you know which root to pick? Do you need to see the solutions for both values of r in order to pick out the correct value of y (the root the two equations have in common)? Or could you have known which of the two roots was correct from only one of the two equations?

Problem 8.6. Solve the equation $y^3 + 36y - 12 = 0$ using the Tschirnhaus method.

Problem 8.7. The Tschirnhaus method amounts to rewriting a cubic polynomial $p(y) = y^3 + py + q$ in terms of a variable $z = y^2 + ry + s$, resulting in a polynomial $q(z) = z^3 - A(r, s)z^2 + B(r, s)z - C(r, s)$, after which an attempt is made to choose r and s so that $A(r, s) = 0$ and $B(r, s) = 0$. We were very fortunate that this technique appears to work perfectly for the cubic. Lagrange, however, realized that this technique would not work in general.

The general strategy is to let $z = y^{n-1} - A_1 y^{n-2} + \cdots + (-1)^{n-2} A_{n-2} y + (-1)^{n-1} A_{n-1}$, and attempt to choose $A_1, \ldots, A_{n-1}$ so that a polynomial $p(y) = y^n - a_1 y^{n-1} + \cdots + (-1)^{n-1} a_{n-1} y + (-1)^n a_n$ becomes $z^n - N$ when rewritten in terms of z.

Consider the case $n = 5$, where $p(y) = y^5 - ay^4 + by^3 - cy^2 + dy - e$. To simplify things, you may assume that $a = 0$, since we know a simple transformation that will bring this about. Let $z = y^4 - py^3 + qy^2 - ry + s$, and show by the same kind of manipulation as in Problem 8.5 that

$$p(y) = y^3 + \left(\frac{pq + c - r}{q - p^2 - b}\right)y^2 + \left(\frac{s - pr - d - z}{q - p^2 - b}\right)y + \frac{e + ps - pz}{q - p^2 - b}.$$

Note that this is a cubic equation in y, so that the equation $p(y) = 0$ can be used to express y in terms of z, just as in the case of the cubic equation. What is the difference between this polynomial and the one that appeared in Problem 8.5? How does this difference complicate the equation satisfied by z after y is expressed in terms of z?

Problem 8.8. Show that the solutions of the cyclotomic (circle-dividing) equation

$$x^4 + x^3 + x^2 + x + 1 = 0,$$

which are the four primitive fifth roots of unity, can *theoretically* be expressed as finite algebraic expressions involving only square and cube roots of rational numbers. In fact, the cube roots turn out to be unnecessary. *You are not being asked to solve the equation. If you try to do so by the Cardano method, you will probably understand why.* The next problem shows the easy route to a realization of this theoretical possibility, following a technique analogous to Viète's solution of the cubic.

Problem 8.9. Use the trigonometric identity

$$\cos(5\theta) = 16\cos^5\theta - 20\cos^3\theta + 5\cos\theta$$

to compute the real part $z = \cos(2\pi/5)$ of a primitive fifth root of unity. Notice that the polynomial $16z^5 - 20z^3 + 5z - 1$ factors as $(z-1)(4z^2 + 2z - 1)^2$, from which it follows that $\cos(2\pi/5)$ is the positive root of the equation $4z^2 + 2z - 1 = 0$. Prove that one of the fifth roots of unity is

$$\frac{-1+\sqrt{5}}{4} + \sqrt{-\frac{5+\sqrt{5}}{8}}.$$

Problem 8.10. Solve the quartic equation $x^4 - 6x^2 - 8x + 24 = 0$, that is,

$$(x^2 + t)^2 = (6 + 2t)\left(x + \frac{4}{6 + 2t}\right)^2,$$

with the other value $t = -4$ that results from solving the resolvent equation.

Problem 8.11. You may have wondered about the factor $r^3 + pr + q$ that factored out of the equation for z when the Tschirnhaus transformation is performed. Can it be zero? If so, what does that fact mean for the equation in y, since this factor is independent of y? Recall that r was chosen (after s was taken as $2p/3$) so as to satisfy the equation $3pr^2 + 9qr - p^2 = 0$. Multiply this equation by r and the equation $r^3 + pr + q = 0$ by $3p$, then subtract to obtain the equation $9qr^2 - 4p^2r - 3pq = 0$. Then eliminate r^2 between this equation and the one above so as to obtain $r(27q^2 + 4p^3) = 0$. Conclude that this other factor equals zero only when the cubic discriminant $p^3/27 + q^2/4$ equals zero; in other words, the original equation has a double or triple root. In that case, the cubic is trivial to solve (see Question 8.1).

Question 8.1. Although the cubic formula for an equation with real coefficients never picks out the double root of the equation (if one exists), that root is always trivial to find without even having to solve the cubic equation. Why? *Hint:* You need calculus to answer this question. Find the derivative of $p(x) = a(x - r)^2(x - s)$.

7. Further reading

Dirk J. Struik, "Ferrari. The biquadratic equation," in *Source Book in Mathematics, 1200–1800,* D. J. Struik, ed., Princeton University Press, Princeton, NJ, 1986.

Dirk J. Struik, "Lagrange. On the general theory of equations," in *Source Book in Mathematics, 1200–1800,* D. J. Struik, ed., Princeton University Press, Princeton, NJ, 1986.

Dirk J. Struik, "Viète. The new algebra," in *Source Book in Mathematics, 1200–1800*," D. J. Struik, ed., Princeton University Press, Princeton, NJ, 1986.

LESSON 9

The Search for Resolvents

It is no exaggeration to say that polynomial equations are nowadays understood fully. There are of course, unanswered questions. In mathematics, there always are. But the work of some brilliant mathematicians over the last 400 years has given us a very complete understanding of what is and is not possible in the formulaic approach to their solution. At the same time, numerical methods have succeeded brilliantly, so that computer algebra programs like *Mathematica, Maple,* and *Matlab* can find the roots of polynomials of even very high degree in a split second.

The story of these advances in knowledge is fascinating, and we shall devote the last three lessons to some of its important moments. In the present lesson, we cover the 150 years from 1620 to 1770. This period begins and ends with major milestones. At the beginning comes the realization that the coefficients of a polynomial are the symmetric functions of its roots. The end is marked by a detailed report on past efforts to solve polynomial equations and a proposal for a unified system that would reduce the solution of a polynomial equation to setting up a system of linear equations in the roots by means of an expression or equation called a *resolvent.*

As we saw in the last chapter, resolvents developed naturally out of the combinatorial approach to the solution of equations. Resolvents in turn naturally focused the attention of mathematicians on permutations of the roots, at first only for the purpose of counting the number of values a function could have when its variables were permuted. That limited goal by itself was sufficient to produce, eventually, the realization that the general quintic cannot be solved algebraically, a topic that forms the subject of Lesson 10. More sophisticated reflections connecting permutations of the roots with the arithmetic operations in the enlarged fields that resulted from adjoining roots led to the construction of a general method—Galois theory—connecting each polynomial equation with a group of permutations in such a way that the algebraic solvability of the equation is perfectly correlated with the algebraic structure of the group. That development will be described in Lesson 11. The subject matter of the present lesson forms the critical bridge between the formal solution of equations, as presented in courses of high-school algebra, and the mathematics taught to seniors and graduate students as modern algebra.

1. Coefficients and roots

The first to call attention to the symmetric functions of the roots (in 1629) was Albert Girard (1595–1632), who called them *factions*. For a set of four numbers $\{r, s, t, u\}$, the factions are $a = r+s+t+u$, $b = rs+rt+ru+st+su+tu$, $c = rst+rsu+rtu+stu$, and $d = rstu$. The equation $x^4 - ax^3 + bx - c = 0$ has roots r, s, t, u. Although these relations seem obvious now, obtained by expanding $(x-r)(x-s)(x-t)(x-u)$, the complications due to coincident roots and complex roots were not fully sorted out at the time. Bombelli's work on complex numbers had at first been taken only as a sort of useful fiction because it enabled mathematicians to make sense of the Cardano formula. Real roots were sought, and the complex numbers were supposed to arise and disappear in between the initial and final stages. Girard listed complex numbers among the roots. He noted, however, that an equation might have fewer roots than its degree, since, as we would now say, two or more roots might be equal.

2. A unified approach to equations of all degrees

What looked like a promising general approach in the work of Tschirnhaus turned out to be less general than had been hoped. Another systematic approach was made by Joseph-Louis Lagrange (1736–1813) in the 1770 survey mentioned in the previous lesson. We have already described this approach in the case of the formulas for solving quadratic and cubic equations, both of which begin by taking the square root of the discriminant and end with a maximally nonsymmetric linear function of the roots—it assumes $n!$ values as the roots are permuted—that can be found from a resolvent equation of lower degree. It was probably the forms

$$x = \frac{a}{2} + z,$$
$$x = \frac{a}{3} + z + Bz^2,$$

in which the solutions of the quadratic and cubic equations can be expressed (see pages 60 and 82) that led Euler to propose in 1732 that the solution of an equation of degree n might be written as

$$x = \sqrt[n]{A_1} + \cdots + \sqrt[n]{A_{n-1}},$$

where $A_1, \ldots, A_{n-1}$ are the roots of a resolvent equation of degree $n-1$.

Thirty years later, he proposed an alternative formula of the same type, namely

$$x = w + A\sqrt[n]{v} + B\sqrt[n]{v^2} + \cdots + Q\sqrt[n]{v^{n-1}}.$$

Such a form for the general solution of the quintic equation was used in Abel's proof that no such formula could exist.

2.1. A resolvent for the cubic equation. To get started on our discussion of resolvents, let us begin with a recap of what we know about the resolvent for the cubic equation. The equation $ax^3 + bx^2 + cx + d = 0$ is equivalent to the non-linear system

$$u + v + w = -\frac{b}{a},$$
$$uv + vw + wu = \frac{c}{a},$$
$$uvw = -\frac{d}{a}.$$

The resolvent technique replaces this non-linear system by an equivalent linear system

$$u + v + w = -b/a,$$
$$u + \alpha v + \alpha^2 w = B,$$
$$u + \alpha^2 v + \alpha w = C.$$

The first of these equations is written down directly from the equation. To find B and C, we use the fact that the function $r(u, v, w) = (u + \alpha^2 v + \alpha w)^3$ assumes only two values when u, v, w are permuted. Those two values are the roots of a quadratic equation whose coefficients are symmetric functions of u, v, w and hence computable in terms of a, b, c, d. From them B (and C also) can then be computed by taking cube roots. Thus the nonsymmetric function $r(u, v, w)$ yields a determinate system of linear equations for the roots. The fact that the values of this function can be found by solving an equation of lower degree and extracting a root, together with the fact that knowing these two values allows us to solve the equation, justifies the term *resolvent* that we have applied to the function $r(u, v, w)$.

This program can be stated for equations of any degree. The pattern that emerges is that the resolvent produced by this approach is found by solving an equation of degree $(n-1)!$ in the variable x^n. The introduction of the roots of unity in this approach guarantees that the resolvent will be an equation of degree $n!$ but will contain only powers of z^n. Solving it amounts to solving an equation of degree $(n-1)!$, then extracting an nth root. That is exactly what happened in the case of the cubic equation. For a quartic equation, we would expect an equation of degree 6 in z^4. Let's give this a try. We don't know how to solve an equation of degree 6, but perhaps we'll find that the resolvent is "sparser" than we thought. It may prove to be a cubic equation in z^8 or a quadratic in z^{12}, if we are lucky.

3. A resolvent for the general quartic equation

Consider the general quartic equation, $x^4 - ax^3 + bx^2 - cx + d = 0$. Assuming that the roots are t, u, v, and w, we follow our previous model, using the fourth roots of unity $\pm i$, ± 1 instead of the cube roots, and set up the system

$$\begin{aligned}
t + u + v + w &= a, \\
t + iu + i^2 v + i^3 w &= A, \\
t + i^2 u + i^4 v + i^6 w &= B, \\
t + i^3 u + i^6 v + i^9 w &= C.
\end{aligned}$$

that is, since $i^2 = -1$, the system

$$\begin{aligned}
t + u + v + w &= a, \\
t + iu - v - iw &= A, \\
t - u + v - w &= B, \\
t - iu - v + iw &= C.
\end{aligned}$$

Thus, A and C can be obtained from each other by interchanging u and w. We need to symmetrize each of A, B, and C in order to get an equation that they satisfy with coefficients that can be expressed in terms of a, b, c, and d. For B this is easy, since $B^2 = ((t + v) - (u + w))^2$ is completely determined by choosing a "companion" to go in the set of parentheses with t, and this can be done in only three ways. In other words, only three different values will result for B^2 when the roots are permuted. Therefore B^2 satisfies the cubic equation

$$\left(z - (t + v - u - w)^2\right)\left(z - (t + u - v - w)^2\right)\left(z - (t + w - u - v)^2\right) = 0,$$

whose coefficients are symmetric functions of t, u, v, w. Because they are symmetric functions of the roots, the coefficients of this equation are expressible in terms of a, b, c, d. Indeed, this equation is the same as the equation

$$z^3 - (3a^2 - 8b)z^2 + (3a^4 - 16a^2 b + 16b^2 + 16ac - 64d)z$$
$$- (a^6 - 8a^4 b + 16a^3 c + 16a^2 b^2 - 64abc + 64c^2) = 0.$$

We have now found that B^2 is a resolvent, and it can be found by solving a cubic equation that we can write down. Now if we are to follow Lagrange's method faithfully, we need another to find some power of A or C that also takes on fewer than 4 values when the roots are permuted. We know exactly how to express it, but solving it turns out to be very discouraging. It is easy to see that A assumes 24 formally different values when the roots are permuted. And even the powers of A are of no help. They all assume at least 6 values when the roots are permuted. Three of the values that A can assume are iA, $i^2 A = -A$, and $i^3 A = -iA$. Thus the polynomial that is the product of all 24 factors $z - \sigma A$, where σ ranges over all permutations of the roots, will break up into a product of six polynomials like $(z - A)(z - iA)(z - i^2 A)(z - i^3 A) = z^4 - A^4$, and will therefore be a polynomial of degree 6 in the variable z^4. However, after the substitution $\xi = z^4$, what results is a *full* equation of degree 6 in ξ, with no zero coefficients. Moreover, those coefficients are horrendously complicated expressions in the coefficients of the original equation. The coefficient of ξ, for example, is a polynomial of degree 20 in t, u, v, w. We need to look for an alternative solution of the original equation.

As it turns out, the answer is already at hand. We have already found three expressions in t, u, v, w, namely $B_1 = t - u + v - w$, $B_2 = t + u - v - w$, and $B_3 = t - u - v + w$. We therefore have a system of four linear equations that we can solve, getting

$$
\begin{aligned}
t &= \frac{a + B_1 + B_2 + B_3}{4}, \\
u &= \frac{a - B_1 + B_2 - B_3}{4}, \\
v &= \frac{a + B_1 - B_2 - B_3}{4}, \\
w &= \frac{a - B_2 - B_2 + B_3}{4}.
\end{aligned}
$$

We have been lucky here. The technique we suggested *led us* to a system of equations that we could solve. But it did not provide easily solvable equations for A and C, which had been our original program. We opportunistically abandoned the linear system that we originally set up and jumped to another that we happened to encounter along the way. The unity of approach we were seeking has therefore failed.

Nevertheless, we have at least discovered a general strategy for solving an equation of any degree n: Look for a nonsymmetric expression in the n roots that assumes *fewer than n values* when the roots are permuted. Form the *resolvent* polynomial whose roots are the values assumed by this expression. (For our purposes, the expression itself or this polynomial or the equation obtained by setting the polynomial equal to zero can all be conveniently referred to as a resolvent.) The resolvent polynomial will be of degree less than n, and its coefficients will be symmetric in the original roots and hence expressible in terms of the coefficients of the original equation. In this way, the symmetry of the original equation will have been broken— perhaps not completely, but at least broken. The challenge then, is to seek such nonsymmetric expressions. The place to begin is with five roots: the quintic equation.

4. The state of polynomial algebra in 1770

Although only partially successful, the method we have just examined suggested a new approach to the general problem of solving equations. For the quintic equation, this approach would work as follows. Starting with the equation itself $x^5 - ax^4 + bx^3 - cx^2 + dx - e = 0$ with roots s, t, u, v, w, which represents the system

$$\begin{aligned}
s + t + u + v + w &= a, \\
s(t + u + v + w) + t(u + v + w) + u(v + w) + vw &= b, \\
st(u + v + w) + su(v + w) + svw + tu(v + w) + tuw + tvw &= c, \\
stuv + stuw + stvw + suvw + tuvw &= d, \\
stuvw &= e,
\end{aligned}$$

replace this system by a linear system with a Vandermonde matrix

$$\begin{aligned}
s + t + u + v + w &= a \\
s + \alpha t + \alpha^2 u + \alpha^3 v + \alpha^4 w &= A \\
s + \alpha^2 t + \alpha^4 u + \alpha v + \alpha 3w &= B \\
s + \alpha^3 t + \alpha u + \alpha^4 v + \alpha^2 w &= C \\
s + \alpha^4 t + \alpha^3 u + \alpha^2 v + \alpha w &= D,
\end{aligned}$$

where $\alpha = \cos(2\pi/5) + i\sin(2\pi/5)$ is a fifth root of unity. The challenge would be to express A, B, C, D in terms of the coefficients of the original equation. That is, we would need to symmetrize these expressions as we have done above and hope for a "sparse" equation containing only powers that are multiples of some multiple of 5. Therein lies the difficulty. To find an expression for the analog of A in the case of the cubic, as we did above, it was necessary to symmetrize, getting an equation of degree $6 = 3!$ in A that contained only powers that were multiples of 3, and hence was quadratic in A^3. For the quartic, we were not so lucky, and were stuck with the equation of degree 6. But perhaps that is because the degree was a composite number. Things may be simpler for equations of prime degree.

To determine A for the quintic equation, we would expect an equation of degree $120 = 5!$ that would be of degree of degree 24 in u^5. Even if it were very "sparse" with only every fourth coefficient nonzero, that would still be an equation of degree 6 in u^{20}, and so our method would fail. Lagrange realized all that, and gave a very pessimistic report on the prospects of finding a general solution of all polynomial equations.

As Lagrange's survey showed, there was no known, systematic way of setting up a system of linear equations for the roots of a general equation of degree n. His use of a Vandermonde matrix formed on the roots of unity provided about the sparsest resolvent equation one could hope for, and it was not sufficiently sparse to render the quintic equation solvable. The only other general approach that had been suggested, the Tschirnhaus technique of making a substitution one degree less than the degree of the equation, also fails for higher-degree equations. Thus, it seems that the general quintic equation presents a formidable barricade.

Lagrange's pessimism on this point turned out to be justified. Within a few years after he wrote, Paolo Ruffini (1765–1822, primarily a physician) had produced a cogent argument that there was no algebraic formula for solving quintic equations. That dénouement in the drama of polynomial algebra forms part of our next lesson. Now, one could easily believe that such a formula is too complicated for people to discover. How one would come to believe that it doesn't exist at all is quite another matter. What led mathematicians to that amazing conclusion? The answer is found in the analysis of resolvents in five roots.

4.1. Seeking a resolvent for the quintic. To find a resolvent for a quintic equation

$$x^5 - ax^4 + bx^3 - cx^2 + dx - e = 0$$

with roots s, t, u, v, w, we need a function $f(s, t, u, v, w)$ that assumes at most four formally different values $f_j(s, t, u, v, w)$, $j = 1, 2, 3, 4$, when the variables (roots) are permuted.

Remark 9.1. In order for these permutations to be applicable, the functions themselves have to be such that the variables can be interchanged. The simplest class that will meet our needs is the class of rational functions $f(s, t, u, v, w)$, and we shall assume that the functions we deal with are rational.

These four different values will then be the roots of a quartic equation

$$(z - f_1)(z - f_2)(z - f_3)(z - f_4) = 0,$$

whose coefficients are symmetric in s, t, u, v, w and hence expressible as functions of a, b, c, d. Because this is a quartic equation, it can be solved, and then we will have four roots $g_j = f_j(s, t, u, v, w)$, $j = 1, 2, 3, 4$, and hence a system of five equations

$$
\begin{aligned}
s + t + u + v + w &= a \\
f_1(s, t, u, v, w) &= g_1(a, b, c, d) \\
f_2(s, t, u, v, w) &= g_2(a, b, c, d) \\
f_3(s, t, u, v, w) &= g_3(a, b, c, d) \\
f_4(s, t, u, v, w) &= g_4(a, b, c, d).
\end{aligned}
$$

If the functions f_j are not "too messy," we might have a better chance of solving this system than the original system. It certainly has more symmetry than the original system, since the last four equations are all essentially the same equation, with permutations of the variables s, t, u, v, w.

As Problem 9.1 below shows, such a technique actually does work for the quartic equation with roots t, u, v, w, using the function $f(t, u, v, w) = tu + vw$. But something goes wrong when we try to do this with five variables, and that failure has interesting consequences, as we shall see in the next lesson.

Lagrange had provided the essential idea that was to lead to a solution of the problem. To get a resolvent, one should look at suitable linear combinations of the roots, then form the simplest polynomial in each of those linear combinations whose coefficients will be symmetric functions of the roots, and hence expressible in terms of the coefficients of the original equation. Thus, the spotlight was turned onto the problem of the symmetries of a function of n variables when the variables are permuted. For each degree n, we need to find an asymmetric function of the roots for which some power assumes fewer than n values when the roots are permuted. For the cubic equation with roots u, v, and w, the function $(u+\alpha v+\alpha^2 w)^3$ assumes only two values. For the quartic with roots t, u, v, and w, the functions $(t+u-v-w)^2$ and $tu+vw$ both assume only three values. That was the clue mathematicians had to work on in order to solve the general problem. Before we take up this final phase of the story, we need to develop the necessary information about permutations.

5. Permutations enter algebra

Permutations and combinations were first studied for mystical reasons by the ancient Hindu mathematicians. The closely related topic of determinants originated in China and Japan in the sixteenth and seventeenth centuries. All this lore was independently rediscovered by Leibniz in the late seventeenth century. It proved its value a century later in the quest for a general method of solving equations. As we have just seen, the problem was to determine how many different values a function could assume when its variables were permuted.

This problem was to lead ultimately to the fundamental notion of a *group*, the core concept of what is called *modern algebra*. Modern algebra is nowadays taught beginning with groups, but the historical connection with equations is never used as a motivation. A group is introduced as a purely abstract object, with permutation groups playing the role of the most important example. One generally has to get to the second or third semester of modern algebra before polynomial equations are mentioned in the same lecture with groups. Even then, the smooth modern highway leading to the solution of the problem bypasses the old gravel road that the pioneers traveled over.

To speak less metaphorically, the intermediate stage of development between the classical problem of solving polynomial equations and the modern solution to that problem using Galois theory involved counting the number of values a function may have when its variables are permuted. That procedure is omitted from modern algebra courses, since a more natural route to the solution has been found. In our final two lessons, we shall describe the old route, now bypassed, and the new highway to the solution.

6. Permutations of the variables in a function

In this section, we consider functions of an unspecified number of variables. To keep the discussion concrete, we shall illustrate it with the case of a function of four variables $f(t, u, v, w)$, but the possibility of more variables should be kept in mind as the discussion proceeds. As we mentioned above, we shall assume that f is a rational function, so that it will make sense to permute the variables.

There are 24 formally different functions that can be formed from a given function $f(t, u, v, w)$ by permuting the variables t, u, v, w, and we could list all of them if we were inclined to do so, starting with $f_0(t, u, v, w) = f(t, u, v, w)$, $f_1(t, u, v, w) = f(u, t, v, w)$, $f_2(t, u, v, w) = f(v, u, t, w)$, and so forth up to $f_{23}(t, u, v, w)$. For particular functions $f(t, u, v, w)$, these 24 functions may not all be formally different. Indeed, if $f(t, u, v, w) = tu + vw$, then only three different values occur among these functions. They fall naturally into three sets of eight, and all the functions in a given set are the same. There may be only one function, as happens, for example, with the function $f(t, u, v, w) = t + u + v + w$, or there may be only two, as in the case of the square root of the discriminant:

$$d_4(t, u, v, w) = (t - u)(t - v)(t - w)(u - v)(u - w)(v - w).$$

Here $d_4(u, t, v, w) = -d_4(t, u, v, w)$.

We think of these functions as being the *composition* of f with various permutations of the variables. Thus, if $\sigma(t, u, v, w) = (u, t, v, w)$, then $f_1 = f \circ \sigma$. We can now leave the functions f in the background and concentrate on the permutations themselves. Every permutation can be written as a sequence of *transpositions* that simply interchange two letters. We shall use the notation (uv) to indicate the transposition of u and v, so that the effect of (uv) is to replace (t, u, v, w) with (t, v, u, w). The representation of a permutation as such a sequence of transpositions is not unique. However, its *parity*, defined to be odd if the number of transpositions is odd and even if the number is even, is the same in all representations.

The easiest way to see that fact is to imagine the letters in a finite string and count the number of *inversions*, that is, the number of pairs of letters that are *not* in alphabetical order. For example, in the sequence c, f, a, b, d, h, g, e, the pairs (c, a), (c, b), (f, a), (f, b), (f, d), (f, e), (h, g), (h, e), and (g, e) are not in alphabetical order, a total of nine inversions. If we interchange a pair, say a and h, the number of inversions will change. However, all inversions involving letters that precede a or follow h will remain inversions. The only changes will be in those that involve the letters between a and h, and of course a and h themselves. As for the letters in between, the pairs that were inverted with respect to a before are now "uninverted" and vice versa. The same is true for h. Thus the change in the number of inversions is $(m - n) + (p - q)$, where m is the number of letters between a and h that were not inverted relative to a, n is the number that were inverted relative to a, p is the number that were not inverted relative to h and q the number that were. In particular, $m + n = p + q = r$, where r is the number of letters between a and h. It follows that the change is $m - (r - m) + p - (r - p) = 2m + 2p - 2r$, which is an even number. Then if we consider the pair (a, h) itself, we see that it was originally not an inversion, but after the interchange, we have (h, a), which *is* an inversion. Thus, when a transposition is performed, the number of inversions changes by an odd number. Therefore, any permutation with an odd number of inversions can be written only as the result of a sequence of an odd number of transpositions, starting from the natural order (which we take to be alphabetical for sets of letters and ascending for sets of numbers), and any with an even number is the result of a sequence of an even number of transpositions.

Every permutation can be classified as even or odd according to the number of transpositions in any representation of it as a finite sequence of transpositions. Obviously an even permutation followed by an even permutation is even, as is an odd permutation followed by an odd permutation. An even permutation followed by an odd one, or an odd one followed by an even one, is odd.

6.1. Two-valued functions. Particular interest attaches to a two-valued function. Suppose that the function $f(t, u, v, w)$ can assume only two possible values. Let these values be $f(t, u, v, w) = f_0(t, u, v, w)$ and $f_1(t, u, v, w)$, which is different from f_0. Notice that if σ and τ are permutations such that $f \circ \sigma = f$ and $f \circ \tau = f$, then $f \circ (\sigma \circ \tau) = (f \circ \sigma) \circ \tau = f \circ \tau = f$. It follows that if f assumes exactly two values, there must be some transposition σ such that $f \circ \sigma = f_1$. (Otherwise f would assume only one value, since every permutation is equivalent to a sequence of transpositions.) Without loss of generality, we may assume that $\sigma = (tu)$. Now we claim that $f_1 \circ \sigma = f$; for if not, then $f_1 \circ \sigma = f_1$, from which it follows that $f_1 = f_1 \circ \sigma = (f \circ \sigma) \circ \sigma = f \circ (\sigma \circ \sigma) = f$, which contradicts the assumption that $f \neq f_1$. Hence there must be *at least* one transposition that interchanges the two values.

It now follows that if $\tau = (tv)$, then $f \circ \tau = f_1$ also; for if not, then $f \circ \tau \circ \sigma = f \circ \sigma = f_1$. If $\nu = \tau \circ \sigma$, we note that $\nu \circ \nu \circ \nu$ is the identity permutation that moves nothing. Hence surely $f \circ (\nu \circ \nu \circ \nu) = f$. Our hypothesis is that $f \circ \nu = f_1$ What can $f \circ \nu \circ \nu$ be? It must be $f_1 \circ \nu$. If this is f, then $f \circ \nu = (f_1 \circ \nu) \circ \nu = f \circ \nu \circ \nu \circ \nu = f$, which contradicts our hypothesis. Thus, if one transposition (tu) interchanges the two values, then all transpositions (tv) (for any v) interchange the two values. But then, for any w, (vw) must also interchange the two values; that is, if one transposition interchanges the two values, then *all* transpositions interchange the two values. Hence $f \circ \sigma = f_1$ if σ is an odd permutation and $f \circ \sigma = f$ if σ is an even permutation.

Now consider the functions $g(t, u, v, w) = \big(f(t, u, v, w) + f_1(t, u, v, w)\big)/2$ and $h(t, u, v, w) = \big(f(t, u, v, w) - f_1(t, u, v, w)\big)/2$. It is clear that g is symmetric, since any transposition leaves it invariant. As such, it can be written as a function $G(a, b, c, d)$ of the coefficients of the polynomial $x^4 - ax^3 + bx^2 - cx + d$ having t, u, v, w as roots.

In contrast, h is antisymmetric in that $h \circ \sigma = -h$ for each transposition. The function
$$s(t, u, v, w) = \frac{h(t, u, v, w)}{d_4(t, u, v, w)},$$

where $d_4(t, u, v, w) = \sqrt{D_4(a, b, c, d)}$ is the square root of the discriminant, is symmetric. Hence $s(t, u, v, w)$ can be expressed as a function H of the coefficients of the polynomial $x^4 - ax^3 + bx^2 - cx + d$ having t, u, v, w as roots. Thus we have

$$h(t, u, v, w) = H(a, b, c, d)d_4(t, u, v, w) = H(a, b, c, d)\sqrt{D_4(a, b, c, d)},$$

where $D_4(a, b, c, d)$ is the quartic discriminant $(t - u)^2(t - v)^2(t - w)^2(u - v)^2(u - w)^2(v - w)^2$. (It suffices to know that *in principle* D_4 can be expressed in terms of the coefficients a, b, c, and d. The actual expression is horrendously complicated and not enlightening.) Putting these expressions together, we see that any two-valued function of the roots can be written in terms of the coefficients as

$$(2) \qquad f(t, u, v, w) = G(a, b, c, d) + H(a, b, c, d)\sqrt{D_4(a, b, c, d)}.$$

An important special case is that of a quadratic equation with the two roots t and u. We can take the function $f(t, u) = t$, which assumes only the two values t and u when the roots are permuted. The formula analogous to (2) is the quadratic formula for solving the equation $x^2 - ax + b = 0$, that is, $G(a, b) = a/2$, $H(a, b) = 1/2$, and $D_2(a, b) = a^2 - 4b$.

Another good example occurs in the case of a cubic equation $y^3 + py + q = 0$ with roots u, v, w. As we have seen, the function $f(u, v, w) = \frac{1}{27}(u + \alpha v + \alpha^2 w)^3$ takes on only two values, and can be expressed as $-q/2 + \sqrt{p^3/27 + q^2/4}$; that is, taking $a = 0$, $b = p$, $c = -q$, we get

$$f(u, v, w) = G(a, b, c) + H(a, b, c)\sqrt{D_3(a, b, c)},$$

where $G(a, b, c) = c/2$, $H(a, b, c) = 1$, and $D_3(a, b, c) = b^3/27 + c^2/4$.

The importance of two-valued functions lies in the following theorem proved by Augustin-Louis Cauchy (1789–1856); we shall state it for the case of five variables (s, t, u, v, w) only: *If a function $f(s, t, u, v, w)$ assumes fewer than five values when the variables are permuted, then it assumes at most two values.*

In other words, there are no three- or four-valued functions of five variables. Putting this fact together with the equation-solving program we have formulated, we see that we would need two independent two-valued functions f and g to generate enough equations to produce a system of five independent equations to determine the roots. Moreover, we know in general what these functions would have to look like (Eq. 2).

7. Problems and questions

Problem 9.1. Although the linear system with a Vandermonde matrix constructed from the roots of unity is by far the simplest strategy for solving a polynomial equation, we have seen that it may not work. The failure of this particular method by no means indicates that there is no formula for solving the equation, as we have seen in the case of the quartic equation. In fact, the solution that we have given in the case of the quartic is more complicated than it needs to be. Consider again the quartic $x^4 - ax^3 + bx^2 - cx + d = 0$ with roots t, u, v, w. Show that the function $tu + vw$ has only three values when the roots are permuted and that these values are the three solutions of the cubic equation

$$z^3 - bz^2 + (ac - 4d)z + (4bd - c^2 - a^2d) = 0.$$

(The algebra will be very tedious. It would be advisable to resort to a computer algebra program to do this verification.)

Let these roots be g_1, g_2, and g_3, and show that the system of equations

$$\begin{aligned}
t + u + v + w &= a, \\
tu + vw &= g_1, \\
tv + uw &= g_2, \\
tw + uv &= g_3,
\end{aligned}$$

implies the linear system

$$\begin{aligned}
t + w &= \frac{a + \sqrt{a^2 - 4(g_1 + g_2)}}{2}, \\
u + v &= \frac{a - \sqrt{a^2 - 4(g_1 + g_2)}}{2}, \\
t + u &= \frac{a + \sqrt{a^2 - 4(g_2 + g_3)}}{2}, \\
v + w &= \frac{a - \sqrt{a^2 - 4(g_2 + g_3)}}{2}.
\end{aligned}$$

Hint: Adding the second and third equations produces the product $(t + w)(u + v)$, and the first equation yields the sum $(t + w) + (u + v)$. Use what you know about finding two quantities from their sum and product. Then do the same with the third and fourth equations.

Problem 9.2. Show that only three of the four equations in the linear system written in the previous problem are independent.

Problem 9.3. Having determined $tu + vw$ by solving the cubic equation in Problem 9.1, and knowing the product $tuvw = d$, show how to determine tu and vw. Then, knowing $t + u$, show how to determine t and u, and likewise v and w. (Thus the one three-valued function we found does suffice to determine all four of the roots.)

Problem 9.4. (*Cycles.*) Given a permutation σ, we can start with any letter, say t and then proceed to the letter that t replaces under the permutation σ, then to the letter replaced by that letter, and so on, until eventually we get back to t. For example, if $\sigma(t, u, v, w) = (u, w, v, t)$, then t replaces w, which replaces u, which replaces t. We write this permutation as (twu). It is called a *3-cycle*, and is said to *have order* 3, meaning that if it is applied three times, the result is the identity permutation, which moves nothing.

A transposition is a 2-cycle. Obviously, every permutation is a unique sequence of disjoint cycles, and two disjoint cycles have the same effect when applied in either order.

List all the different types of permutations of five symbols, that is, a single 2-cycle, a pair of disjoint 2-cycles, a single 3-cycle, and so on.

Problem 9.5. (*The order of a permutation.*) We use the notation $\sigma^2 = \sigma \circ \sigma$ and so forth. Since there are only $n!$ permutations of n symbols, two of these powers must be equal, that is, $\sigma^p = \sigma^q$ for some $q > p$. Then σ^{q-p} must be the identity permutation. The smallest positive value of $q - p$ for which that relation holds is called the *order* of the permutation. For example, the order of a transposition is 2.

Show that the order of a sequence of disjoint cycles is the least common multiple of the orders of the cycles in the sequence.

Problem 9.6. What are the possible orders of a permutation of four symbols?

Problem 9.7. Show that the order of a permutation of five symbols must be 1, 2, 3, 4, 5, or 6. Which ones have order 5?

Problem 9.8. Suppose that the function $f(s, t, u, v, w)$ assumes fewer than five values when its arguments are permuted. Let σ be any 5-cycle. Show first that σ^5 is the identity, and hence that $f \circ \sigma^5 = f$. Then show that there must be two distinct nonnegative integers i, j, $1 \le i < j \le 5$, such that $f \circ \sigma^i = f \circ \sigma^j$, and hence $f \circ \sigma^{j-i} = f$. Show that $f \circ \sigma = f$, and hence $f \circ \sigma^k = f$ for all k. If $j - i = 1$, you are done. If $j - i = 2$, show that $f \circ \sigma^4 = f$ and therefore $f = f \circ \sigma^5 = (f \circ \sigma^4) \circ \sigma = f \circ \sigma$. Give a similar argument if $j - i = 3$ or $j - i = 4$. Conclude that f is invariant under all 5-cycles.

Problem 9.9. Show that the 5-cycle $(suwtv)$ followed by the 5-cycle $(wuvts)$ has the same effect as the 3-cycle (stu). Conclude that if f is invariant under all 5-cycles, it is also invariant under all 3-cycles.

Problem 9.10. Suppose f is invariant under all 5-cycles and $f \circ (st) = f_1$. Show that $f \circ (tu) = f_1$ also by considering the equation $f \circ (st) \circ (tu) = f \circ (stu) = f$. Conclude that if f assumes fewer than five values, then it assumes either exactly one value under transpositions (it is symmetric) or exactly two values, as stated above.

Problem 9.11. Suppose that f is invariant under 3-cycles. Write a 4-cycle $(stuv)$ as $(stu)(uv)$, and deduce that $f \circ \sigma^j$ assumes at most two values for any four-cycle σ as j ranges over the positive integers: $j = 1, 2, \ldots$.

Problem 9.12. A once-popular puzzle is shown in Fig. 11 (a). The numbers are on small squares of wood or plastic with grooves on the left and bottom sides and tongues on the right and top sides that fit those grooves. The frame around the outside has grooves on the top and right sides and tongues on

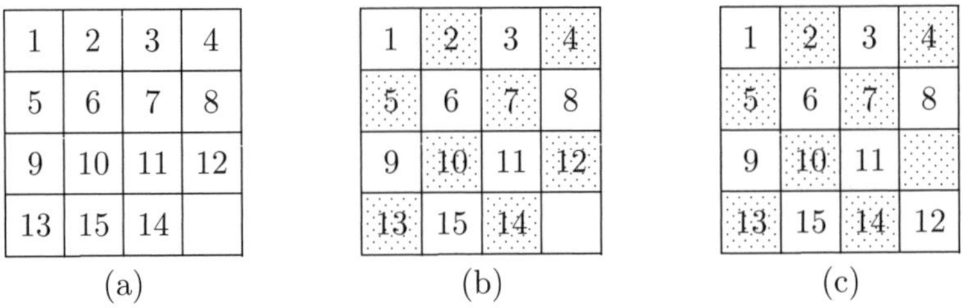

1	2	3	4
5	6	7	8
9	10	11	12
13	15	14	

(a)

1	2	3	4
5	6	7	8
9	10	11	12
13	15	14	

(b)

1	2	3	4
5	6	7	8
9	10	11	
13	15	14	12

(c)

FIGURE 11. A sliding-frame numbers puzzle.

the bottom and left, so that the squares can be slid up or down into the one empty space.

In effect, if you imagine that the empty space bears the number 16, you can always transpose "16" with whatever number is immediately above, below, left, or right. Notice that when the puzzle comes to you, the numbers 14 and 15 are out of order. The challenge is to get all 15 numbers in the correct order, leaving the blank space at the lower right corner.

In its initial configuration, the puzzle has precisely one inversion, and you need to get it so that there are no inverted pairs. Show that this is impossible.

Hint: You need to impose a second "checkerboard" structure on the puzzle. Show that, no matter how the squares are moved around, the total number of inversions is odd if "16" is on an unshaded square.

Problem 9.13. Repeat Problem 9.12, this time with a 5×5 puzzle, in which all numbers are in the correct order except 23 and 24, which are reversed. You can skip the checkerboard structure this time. Why?

Question 9.1. Suppose that $f(s, t, u, v, w)$ assumes exactly two values under transpositions. Can it assume more than two values under other permutations of the variables?

Question 9.2. Is there a function $f(s, t, u, v, w)$ that assumes 120 different values under permutations of its arguments? *Hint:* Think of Lagrange's example for the cubic equation.)

Question 9.3. How might Lagrange have come to discover the expression $\frac{1}{3}(u + \alpha v + \alpha^2 w)$, used to construct the resolvent equation for the cubic? Could it have been an extrapolation from the case of the quadratic equation? As a different possibility, recall also that, because of the work of Bombelli, he knew that the three roots u, v, and w of the cubic equation $y^3 + py + q = 0$ are given by

$$
\begin{aligned}
u &= a + b, \\
v &= \alpha a + \alpha^2 b, \\
w &= \alpha^2 a + \alpha b,
\end{aligned}
$$

where $\alpha = -1/2 + (\sqrt{3}/2)i$ is a cube root of unity and

$$a = \sqrt[3]{-\frac{q}{2} + \sqrt{\frac{q^2}{4} + \frac{p^3}{27}}},$$

$$b = \sqrt[3]{-\frac{q}{2} - \sqrt{\frac{q^2}{4} + \frac{p^3}{27}}}.$$

Solve the first two of these equations for a and b, using the identity

$$\frac{1}{\alpha^2 - \alpha} = \frac{1}{3}(\alpha - \alpha^2).$$

8. Further reading

Edgar Dehn, *Algebraic Equations. An Introduction to the Theories of Lagrange and Galois*, Dover, New York, 1930.

Klaus Mainzer, "Symmetries in mathematics," in *Companion Encyclopedia of the History and Philosophy of the Mathematical Sciences*, I. Grattan-Guinness, ed., Vol. 2, London, Routledge, 1994.

Luboš Nový, *Origins of Modern Algebra*, Noordhoff International Publishing, Leyden, 1973.

Dirk J. Struik, "Girard. The fundamental theorem of algebra," in *Source Book in Mathematics, 1200–1800*," D. J. Struik, ed., Princeton University Press, Princeton, NJ, 1986.

Dirk J. Struik, "Lagrange. On the general theory of equations," in *Source Book in Mathematics, 1200–1800*," D. J. Struik, ed., Princeton University Press, Princeton, NJ, 1986.

Part 4

Abstract Algebra

The search for resolvents for the quintic equation led mathematicians to examine the properties of permutations. Although permutations form part of combinatorics and are one of the oldest parts of mathematics, having been studied in India for more than two thousand years, their reappearance in the nineteenth century in the context of polynomial equations resulted in an explosion of generalization in all areas of mathematics. In short order, this new abstraction finished off attempts to find algebraic formulas to solve equations of degree higher than four and gave definitive answers to some old geometric problems that the ancient Greeks had studied. That is where we end our story, right on the threshold of a new world of thought: Modern algebra, and its siblings modern analysis and geometry, all of which are steeped in the abstract algebraic structures that grew out of this systematic study of permutations.

Existence and Constructibility of Roots

In the five and a half decades from 1770 to 1825, answers to some of the mysteries of polynomial algebra were achieved in the work of several mathematicians, notably Paolo Ruffini, Augustin-Louis Cauchy, Carl Friedrich Gauss (1777–1855), and Niels Henrik Abel (1802–1829). The two major events were the following:

1. The first proof by Gauss in 1799 that the complex numbers are algebraically closed.
2. A convincing argument by Ruffini, also in 1799, that the general quintic equation cannot be solved by a single algebraic formula in terms of the coefficients. Ruffini's proof was endorsed and elaborated by Cauchy 15 years later. A second proof was offered by Abel a decade after Cauchy's work.

1. Proof that the complex numbers are algebraically closed

The recognition that a complex number has an nth root (in fact, n of them), meant that not only could the four arithmetical operations be performed in the complex numbers; roots could be extracted as well. Thus, if there were some algorithm for solving every equation using only arithmetical operations and root extractions (an algebraic method of solving every equation), it would follow, since these operations do not require any new numbers beyond the complex numbers, that every equation with complex coefficients has a root in the complex numbers. Notice, however, that the converse could well be false. It might be that there *exists* a solution of every polynomial equation in the complex numbers, and yet it may be impossible to express that root in terms of the coefficients using only a finite number of algebraic operations. Actually, that just happens to be the case! Gauss suspected as much, but he proved only the first part. He did, however, distinguish between the abstract *existence* of the root, which he proved, and the existence of an algebraic method of computing it starting from the coefficients, which he doubted.

The fact that the complex numbers are closed used to be called the "fundamental theorem of algebra." Actually, the complex numbers belong more to analysis and geometry than to algebra, and the theorem is not at all the basis of algebra. It is an easy theorem to prove using the theory of analytic functions of a complex variable, but a purely algebraic proof is a different matter. The complex numbers are built up from the real numbers, which are an essentially geometric structure, as opposed to the rational numbers, which develop naturally out of ordinary counting with integers from purely arithmetic considerations. To get from the rational numbers to the real numbers, it is essential to introduce some concept of a limit involving sequences of rational numbers or least upper bounds. These limiting procedures lie at the heart of the mathematical notions of continuity and connectedness, which are essentially topological notions. The complex numbers can be constructed by purely algebraic processes starting from the real numbers, but the real numbers cannot be so constructed starting from the rational numbers. Any proof that the complex numbers are algebraically closed must explicitly or implicitly make some use of geometric or topological ideas. Without going into all the details, we will show briefly why it suffices to prove this theorem for polynomials with real coefficients and sketch the principles on which the first proof, given by Gauss in 1799, was based.

Suppose $p(z) = z^n + a_1 z^{n-1} + \cdots + a_{n-1}z + a_n$. Here, $a_1, \ldots, a_n$ are complex numbers, say $a_k = u_k + iv_k$, where u_k and v_k are real numbers. We recall that the *complex conjugate* of a_k (the number $u_k - iv_k$) is denoted $\bar{a}_k$. Notice that $\bar{a}_k a_k = u_k^2 + v_k^2$, which is a nonnegative real number, and $\bar{a}_k + a_k = 2u_k$ is also a real number. Form the polynomial $q(z) = \overline{p(\bar{z})} = z^n + \bar{a}_1 z^{n-1} + \cdots + \bar{a}_{n-1}z + \bar{a}_n$. It is easy to see that if z is a root of p, then $\bar{z}$ is a root of q, and vice versa. Now the polynomial $r(z) = p(z)q(z)$, which is of degree $2n$, has *real coefficients*. For example, if $k < n$, then the coefficient of z^k in $r(z)$ is $\bar{a}_k + \bar{a}_{k-1}a_1 + \cdots + \bar{a}_1 a_{k-1} + a_k$. You can see by pairing terms in this expression that this coefficient is real. If the theorem is proved for polynomials with real coefficients, it follows that $r(z) = 0$ for some value of z, and hence either $p(z) = 0$ or $p(\bar{z}) = 0$. This reduction is needed for the first of the four proofs that Gauss gave.

When the coefficients are real, the equation $p(z) = P(x, y) + iQ(x, y) = 0$, where $z = x + iy$, can be written as a pair of simultaneous equations in x and y: $P(x, y) = 0$ and $Q(x, y) = 0$. For example, a quadratic equation $az^2 + bz + c = 0$ becomes the two real equations $ax^2 + bx - ay^2 + c = 0$, which is a hyperbola, and $(2ax + b)y = 0$, which represents the union of the x-axis ($y = 0$) and the vertical line $x = -b/2a$. (The two intersecting lines that represent this second equation are considered a degenerate hyperbola.) These curves do intersect. If the intersection occurs where $y = 0$, then the real part x satisfies the original equation. If the quadratic discriminant is negative, the hyperbola and the vertical line intersect at the two vertices of the hyperbola, which have coordinates $x = -b/2a$, $y = \pm\sqrt{c/a - b^2/(4a^2)}$. Gauss was able to show by topological considerations that the curves represented by these two polynomial equations in two variables would intersect in this way for any positive degree of the original equation, in fact, that they would intersect in n points.

His argument was that $p(z)$ can be written as

$$p(z) = z^n\left(1 + \frac{a_1 z^{n-1} + \cdots + a_{n-1}z + a_n}{z^n}\right).$$

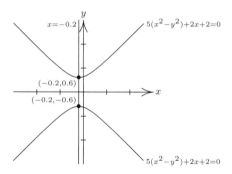

FIGURE 12. Graphical solution of $5z^2 + 2z + 2 = 0$ taking $z = x + iy$.

He noted that if $z = r(\cos\theta + i\sin\theta)$ and r is very large, this last expression shows that the points on the circle $|z| = r$ where the real part of $p(z)$ is zero will be very close to the points where the real part of z^n is zero, since $p(z) = z^n(1 + \varepsilon)$, where ε is very small. (This is a continuity argument.) But we know exactly where the real part of z^n is zero. It vanishes at the points $r(\cos\theta + i\sin\theta)$ where $\cos(n\theta) = 0$, that is $\theta = \frac{(k+\frac{1}{2})\pi}{n}$, $k = 1,\ldots,n$. The curve $P(x, y) = 0$ representing the points where the real part of $p(z)$ vanishes must be a curve that wanders through points very near to these n points. Similarly, the points where the imaginary part is zero must form a curve $Q(x, y) = 0$ that wanders through points very close to the points where $\sin(n\theta) = 0$, that is, the points $r(\cos\theta + i\sin\theta)$ where $\theta = k\pi/n$, $k = 1,\ldots,n$. Since the points in this last set alternate with the points in the first set as we traverse the circle $|z| = r$, the two curves cannot weave through these two sets of points without intersecting n times.

Example 10.1. We illustrate how the two curves intersect in the case of the equation $5z^2 + 2z + 2 = 0$. We let $z = x + iy$ and set the real and imaginary parts of the polynomial equal to zero. That is, we form the system

$$P(x, y) = 5(x^2 - y^2) + 2x + 2 = 0,$$
$$Q(x, y) = 10xy + 2y = 0.$$

The equation $P(x, y) = 0$ describes the hyperbola shown in Fig. 12, and the equation $Q(x, y) = 0$ amounts to $y(5x + 1) = 0$, which is a degenerate hyperbola consisting of the horizontal line $y = 0$ (the x axis) and the vertical line $x = -0.2$. These two curves intersect in two points representing the solutions of the equation, namely $z = -0.2 + 0.6i$ and $z = -0.2 - 0.6i$.

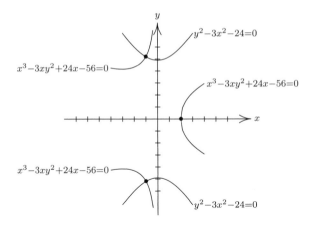

FIGURE 13. Graphical solution of $z^3 + 24z - 56 = 0$ taking $z = x + iy$.

Example 10.2. Consider the cubic equation $z^3 + 24z - 56 = 0$. If $z = x + iy$, this equation becomes the system

$$P(x, y) = x^3 - 3xy^2 + 24x - 56 = 0,$$
$$Q(x, y) = y(3x^2 - y^2 + 24) = 0.$$

In the portion of the complex plane shown in Fig. 13, the first of these curves consists of several disconnected pieces. The second is the union of the x axis and the hyperbola $3x^2 - y^2 + 24 = 0$. The curves intersect in the three points corresponding to the roots $2 + 0i$, $-1 + 3\sqrt{3}i$, and $-1 - 3\sqrt{3}i$, as shown in Fig. 13.

2. Solution by radicals: General considerations

In this section we tie together what we know about lower-degree equations as background for our discussion of Abel's 1826 revision of his 1824 proof of the non-existence of an algebraic formula for solving the general quintic equation

$$x^5 - ax^4 + bx^3 - cx^2 + dx - e = 0.$$

The argument proceeds by showing that certain properties possessed by the quadratic and cubic formulas are properties that any general formula must have, then showing that no formula for the quintic can possibly have them. To state these properties, we shall yet again repeat some facts we have derived in earlier lessons.

2.1. The quadratic formula. As we saw in Lesson 6, the quadratic formula for solving the equation $x^2 - ax + b = 0$ can be written as

$$x = x(a, b) = \frac{a}{2} + z,$$

where $z = \sqrt{a^2/4 - b}$. Here we have omitted the $\pm$ sign usually included in this formula. As you know, the square root sign is ambiguous. When complex numbers are substituted for the variables a and b, z may represent either of two numbers, except in the unusual case when $b = a^2/4$. However, if we substitute $a/2 + z$ for x in the polynomial $x^2 - ax + b$ and replace z^2 by $a^2/4 - b$, the result is

$$(a - a)z + \left(\frac{a^2}{4} - b + \frac{a^2}{4} - \frac{a^2}{2} + b\right).$$

Elementary algebra shows that both the coefficient of z and the term independent of z are formally, identically zero as functions of a and b. The only fact we needed to get this relation was the equation $z^2 = a^2/4 - b$. Since this equation involves only z^2, not z, we do not have to worry about the ambiguity of the square root, and hence could omit the $\pm$ sign that is used when the letters a and b are replaced by specific numbers. If the two roots of the polynomial are u and v, then $a = u + v$ and $b = uv$, so that $a^2/4 - b = (u - v)^2/4$. Thus $z = (u - v)/2$ or $(v - u)/2$, so that z may be either of two *polynomials* when expressed in terms of the roots.

Remark 10.1. The ambiguity of the square root requires some explanation, since it is one of the sources of an annoying vagueness in Abel's reasoning. We take for granted, as Abel did, that we are dealing with complex numbers, and that each complex number except 0 has two square roots in the complex plane. The ambiguity in the equation $w = \sqrt{z}$ can be removed by giving a precise specification of the square root that we want. For every complex number z except 0, we can write $z = r(\cos\theta + i\sin\theta)$ in a unique way if we specify $r > 0$ and $0 \leq \theta < 2\pi$. Then $\sqrt{z}$ can be defined unambiguously either as $\sqrt{r}\left(\cos(\theta/2) + i\sin(\theta/2)\right)$ or as $-\sqrt{r}\left(\cos(\theta/2) + i\sin(\theta/2)\right) = \sqrt{r}\left(\cos(\pi + \theta/2) + i\sin(\pi + \theta/2)\right)$, where $\sqrt{r}$ is made unambiguous by requiring it to be the positive square root of r. Let us call the first of these square roots w_1 and the second w_2. In both cases, $w_1^2 = z = w_2^2$, and we may choose to think of them as two *different* mappings $z \mapsto w$. For that reason, we might prefer to think of two different z values z_1 and z_2 occupying parallel planes, as shown in Fig. 14. Cauchy and Victor Puiseux (1820–1883) used this way of keeping track of the different "sheets" of an algebraic function.

The subscripts we have attached to z here help us keep track of what is really going on when we write $w = \sqrt{z}$, but they do not really capture all the nuances of the situation. We need to overcome the asymmetry between the z's and the w's: w_1 and w_2 inhabit the same plane, while z_1 and z_2 "live" in parallel universes. A variable point in the plane of the two w's can move smoothly from w_1's territory to w_2's by simply crossing the real line. The corresponding z's would have to leap across empty space to do that. Moreover, if we imagine a variable point z starting at $z = r$ on the positive real axis in the z_1 plane and moving counterclockwise around a circle centered at the origin, as it approaches the starting point from the lower half-plane, w_1 will not be approaching its starting value. Rather, it will be approaching the starting value of w_2. We wish to take account of the significant fact that when z_1 approaches its starting point, w_1 tends to the starting point of w_2. This strongly suggests that z_1 should morph into z_2 when it crosses its positive real axis, and vice versa. Such an idea, inspired by an 1850 paper of Puiseux, was developed by Bernhard Riemann (1826–1866) the following year.

The technique for bringing about a smooth transition from z_1 to z_2 is intuitive and is illustrated in Fig. 14. The two z-planes are cut along the positive real axis or any other ray emanating from the "branch point" 0. Then the lower side of each cut is glued to the upper side of the other. (This will be easiest to visualize if you imagine the z_2-plane picked up and turned over so that the dotted edge of the z_2-plane lies on the dotted edge of the z_1 plane.) The result is the *Riemann* surface of the function $w = \sqrt{z}$. It consists of two "sheets" (copies of the complex plane) glued together as just stated. You can easily make a model of this surface with two sheets of paper, a pair of scissors, and cellophane tape. On such a model you can move your finger smoothly and continuously over the entire Riemann surface, without any jumps when it moves from the z_1 sheet to the z_2 sheet. In particular, if you describe a small circle about the branch point 0 at the end of the cut, you will see that it crosses over to the back of the paper when it moves across the dotted edges that have been glued together, makes a whole circle on the back, then crosses over again to the front when it moves across the solid line.

At every point on the Riemann surface except the branch point $z = 0$, the mapping $z \mapsto w$ is *analytic*, that is, it has a power-series representation. For example, near the point $z_1 = 2$, we can express w as a series of powers of $z - 2$ using the binomial theorem:

$$\sqrt{z} = \sqrt{2 + z - 2} = \sqrt{2}\sqrt{1 + (z-2)/2}$$

$$= \sqrt{2}\Big(1 + \frac{1}{2}(z-2) + \frac{\frac{1}{2}(-\frac{1}{2})}{2^2 2!}(z-2)^2 + \frac{\frac{1}{2}(-\frac{1}{2})(-\frac{3}{2})}{2^3 3!}(z-2)^3 + \cdots\Big)$$

$$= \sqrt{2}\Big(1 + \frac{1}{2}(z-2) - \frac{1}{32}(z-2)^2 + \frac{1}{128}(z-2)^3 - \frac{5}{2048}(z-2)^4 + \cdots\Big).$$

The portion of the infinite series shown here yields the approximation $\sqrt{2.3 + 0.5i} \approx 1.52539 + 0.16396i$. The computer value of this number is $1.5254 + 0.163891i$. Similarly, from the five terms of the series we have shown, we find $\sqrt{2.2 - 0.6i} \approx 1.49661 - 0.200465i$, while the computer value is $1.49672 - 0.200438i$. Notice that the first value yielded by the series was a w_1-value, lying in the upper half-plane, while the second was a w_2-value, lying in the lower half-plane. That is consistent with the way we glued the two sheets together, since the point $z_1 = 2$ has z_1-values above it and z_2-values below it.

On the other hand, the function $w = \sqrt{z^2}$ is really two separate functions $w = z$ and $w = -z$, and it is not possible to move z around in such a way that one of these values switches into the other, since when z traverses a

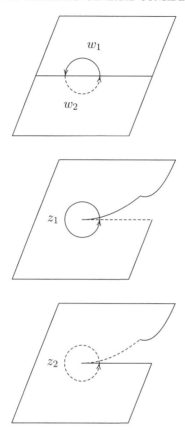

FIGURE 14. The Riemann surface of $w = \sqrt{z}$. The solid z_1 circle corresponds to the solid w_1 semicircle, and the dashed z_2 circle to the dashed w_2 semicircle.

full circle, w does also. The two possible values of w do not share a single plane. Thus, we have the peculiar situation that the quadratic formula for the roots of a polynomial $x^2 - ax + b$ defines two different functions (the two roots) when expressed in terms of the roots u and v, but a single algebraic function having a two-sheeted Riemann surface when expressed in terms of the coefficients a and b. Of course, the situation is more complicated than we have described, since we are now dealing with functions of more than one complex variable.

The example of $\sqrt{z}$ illustrates two important points about the behavior of algebraic functions in general. First, they are generally multi-valued functions, but it is possible to "navigate" from one value to another without any abrupt jumps by moving around on a Riemann surface. Second, the coefficients of the power series representation can be computed from the values of the function along any arc, no matter how short. If the function takes on only the value zero along such an arc, all those coefficients will be zero, and hence the function will be identically zero. In the present context, that means that if an algebraic function $f(a, b, c, d, e)$ produces one root of a polynomial $p(x) = x^5 - ax^4 + bx^3 - cx^2 + dx - e$, that is, $p(f(a, b, c, d, e)) \equiv 0$ at all points near one point, then this relation holds identically at all points (a, b, c, d, e). But, as the example of $w = \sqrt{z}$ shows, it is possible that if the point (a, b, c, d, e) is moved around, the root $(f(a, b, c, d, e)$ may be a different one when it returns to its starting point. In fact, it is guaranteed that starting from any point $(a_0, b_0, c_0, d_0, e_0)$ and any value of $f(a_0, b_0, c_0, d_0, e_0)$, it is possible to vary (a, b, c, d, e) continuously in such a way as that $f(a_0, b_0, c_0, d_0, e_0)$ takes on every possible root when (a, b, c, d, e) passes through the point $(a_0, b_0, c_0, d_0, e_0)$.

This pleasant situation comes about because Riemann surfaces are connected. It is possible for a point to traverse an arc from any point to any other point. If the power series expansion of a function is identically zero at the starting point, it must remain so. The principle just described is sometimes called the *permanence of functional relations* in the theory of functions of a complex variable. It allows us to ignore the multivaluedness of the radicals in a formula and treat the formula just like a single-valued function.

Unfortunately, Abel did not have the Riemann surface at his disposal, since Riemann was born in the year when he published his paper on the insolvability of the quintic equation. As a result, his arguments suffer from a certain lack of clarity. In order to understand them, it is necessary to keep the permanence of functional relations in mind. Abel took this principle for granted when working with expressions representing algebraic formulas. As long as he used only the relation $w^n = z$ in his arguments, the ambiguity of the formula $w = \sqrt[n]{z}$ did not matter.

The reason we have made this brief excursion into algebraic geometry is to clarify what Abel took for granted: Two multi-valued algebraic functions *cannot be equal*, except at isolated points, if they assume different numbers of values. For if they are equal along any arc, no matter how short, then they must be identically equal, and that cannot be if they assume different numbers of values. To illustrate by our simple example, a function whose Riemann surface is two-sheeted, like $w = \sqrt{z}$, cannot be equal to a function whose Riemann surface is three-sheeted, like $w = \sqrt[3]{z}$.

2.2. The cubic formula. In Lesson 7, we showed that the way to find the roots of the cubic polynomial $x^3 - ax^2 + bx - c$ was to make the substitution $x = y + a/3$, leading to the "standardized" cubic polynomial $y^3 + py + q$, with $p = b - a^2/3$ and $q = -2a^3/27 + ab/3 - c$. For this polynomial the formula

$$\frac{p}{3(q/2 + \sqrt{q^2/4 + p^3/27})}z^2 + z + \frac{a}{3},$$

where

$$z = \sqrt[3]{-\frac{q}{2} - \sqrt{\frac{q^2}{4} + \frac{p^3}{27}}},$$

when substituted for x in the polynomial $x^3 - ax^2 + bx - c$, yields the function

$$\left(\frac{-p^2}{3\left(q/2 + \sqrt{q^2/4 + p^3/27}\right) + 27q} + \frac{p^2}{3\left(q/2 + \sqrt{q^2/4 + p^3/27}\right)}\right)z^2 + (p-p)z$$

$$+ \left(\frac{p^3 - 27\left(q/2 + \sqrt{q^2/4 + p^3/27}\right)^2 + 27q\left(q/2 + \sqrt{q^2/4 + p^3/27}\right)}{27\left(q/2 + \sqrt{q^2/4 + p^3/27}\right)}\right).$$

The only property of z used in deriving this last relation was the equation $z^3 = -q/2 - \sqrt{q^2/4 + p^3/27}$. Just as in the case of the quadratic equation, the coefficients of z^2 and z and the term independent of z are all identically, formally equal to 0 as expressions in p and q, and hence also as expressions in a, b, and c. It follows that when numerical values are assigned to a, b, and c, each of the three possible values of z will yield this same (identically zero) function of z, and so the result is always a root of the polynomial $x^3 - ax^2 + bx - c$.

Moreover, as with the quadratic, z may be expressed as a polynomial in the roots u, v, and w of the original polynomial:

$$z = \frac{2w - u - v}{6} + i\frac{v - u}{2\sqrt{3}} = \frac{1}{3}w - \left(\frac{1}{6} + \frac{i}{2\sqrt{3}}\right)u - \left(\frac{1}{6} - \frac{i}{2\sqrt{3}}\right)v$$

$$= \frac{1}{3}(\alpha_1 u + \alpha_2 v + \alpha_3 w),$$

where $\alpha_1 = -1/2 - (\sqrt{3}/2)i$, $\alpha_2 = -1/2 + (\sqrt{3}/2)i$, and $\alpha_3 = 1$ are the three cube roots of 1. Actually, z may be any one of six different polynomials in the roots, all of which are obtained by permuting the roots in any one of them. The examples of the quadratic and cubic formulas illustrate some general features of any formula for finding the roots of a polynomial:

1. The formula is a multi-valued function when expressed in terms of the coefficients, that is, it contains radicals.
2. Each radical in the formula is a polynomial when expressed in terms of the roots.
3. The different values the formula can assume when different choices are made for the values of the radicals in it correspond to permutations of the roots of the polynomial.

This correspondence between the allowable values of a radical in the coefficients and the permutations of the roots lies at the heart of the general problem of solving equations algebraically.

2.3. Algebraic functions and algebraic formulas. The functions that we have displayed here as solutions of the general quadratic and cubic equations are called *algebraic* functions of their variables, in analogy with the algebraic numbers that we discussed in Lesson 1. Each function that we have exhibited is expressible using a finite number of rational operations and root extractions. We warned the reader in Lesson 1 that not all algebraic numbers can be generated from the integers or rational numbers in this way, and the same is true of algebraic functions.

Any function $f(a, b, c, d, e)$ such that $p(a, b, c, d, e, f(a, b, c, d, e)) \equiv 0$, where $p(a, b, c, d, e, f)$ is a nonzero polynomial in its six variables, is by definition an algebraic function of its five variables. Thus, a root of the general quintic equation is an algebraic function of the coefficients. But there is a difference between an *implicitly expressed* algebraic *function* and an *explicitly expressed* algebraic *formula*. This particular algebraic function is not expressible as an algebraic *formula* using only a finite number of algebraic operations on the coefficients. That is the content of Abel's theorem, which we are about to explore.

Remark 10.2. The significance of Abel's theorem is seen most clearly against the background just described. There are two important aspects of this background. (1) In any mathematical theory, the more explicit one can be, the better. A vague description is never as good as an explicit *name*. Instead of writing "the number whose cube is 1331," it is far better to write simply "11." (2) Explicitness is relative to the language in which things are expressed. Consider, for example, the following description of the ellipse by Apollonius, which most people will find bewildering: *It is a curve such that the square of the ordinate from any point to the axis equals the rectangle applied to the portion of the axis cut off by the ordinate and whose defect on the axis is similar to the rectangle contained by the axis and the latus rectum.* (Apollonius gave an extremely complicated description of the *latus rectum*, or *upright side*, *within* this definition!) Now consider the definition given in modern calculus books: *An ellipse is the locus of a point moving in such a way that the sum of its distances from two fixed points is constant.* Or, even better, *An ellipse is a curve whose equation is $x^2/a^2 + y^2/b^2 = 1$.* Apollonius was writing in the language of Euclidean constructions. In that language, the explicit description of an ellipse is complicated. In ours, it can be understood at a glance. This second point is relevant to our current discussion, since the language in which explicitness was being sought for solutions of equations was the language of algebraic formulas. Such a formula serves as a *name* for a number. It turns out that most algebraic numbers do not have names in that language. And, as we shall see in the next chapter, some of those that do have names, like the solution to the general cubic, require complicated circumlocutions analogous to Apollonius' description of the ellipse.

Thus, one moral to be drawn from the argument of Abel that we are about to consider and the Galois theory discussed in the next lesson is that mathematicians were "barking up the wrong tree" in seeking solutions by radicals. Transcendental solutions, such as Viète's, may be far more practical, and may even lead to theoretical advances.

The second moral is even more important: Guided by natural associations and the historical development of the subject, mathematicians set themselves a problem to solve. The eventual solution via Galois theory showed that the problem was not solvable; and even where it was solvable, the solution was not the practical method that the early algebraists would have envisioned. Nevertheless, *the study of this problem was of immense value because of the discoveries that were made while seeking a solution.* Many examples of this phenomenon can be cited in mathematics. For example, although the proof of Fermat's Last Theorem a decade ago merely confirmed what people had always believed during the 350 years when they were trying to prove the theorem, the algebraic number theory that was produced along the way was, like Galois theory, a magnificent triumph of human thought, producing volumes of profound mathematics.

Remark 10.3. We mention in passing that there are nonalgebraic (transcendental) functions, the most elementary of which are exponential functions like 2^x and trigonometric functions like $\sin x$. There is no nonzero polynomial $p(x, y)$ such that $p(x, 2^x) \equiv 0$ or $p(x, \sin x) \equiv 0$. Since we have used trigonometric functions to solve the cubic equation, we know that transcendental formulas can express algebraic functions. That may seem odd, and it could have been avoided in the case of the cubic. But it will turn out to be the only option for expressing the roots of the general quintic polynomial.

3. Abel's proof

Suppose there is an algebraic formula $f(a, b, c, d, e)$ that is a formal root of a polynomial $p(x) = x^5 - ax^4 + bx^3 - cx^2 + dx - e$. When $f(a, b, c, d, e)$ is substituted for x in this polynomial, the resulting function $P(a, b, c, d, e) = p(f(a, b, c, d, e))$ is formally zero. That implies that when any numbers are substituted for the variables a, b, c, d, e, the number $f(a, b, c, d, e)$ will be a root of the corresponding polynomial $p(x)$. Abel derived a contradiction from this assumption, and we are at last ready to say what it was. But a word of warning: The details of Abel's argument are difficult to make clear and precise. Several modern mathematicians have produced arguments in the language of Galois theory that parallel the argument of Abel. These arguments are clear, but would have required considerable background instruction if one were trying to explain them to Abel himself.

Abel wrote the hypothetical formula as

$$(3) \qquad f(a, b, c, d, e) = p_0 + p_1 R^{1/m} + \cdots + p_{m-1} R^{(m-1)/m},$$

where p_k are rational functions of algebraic functions of a, b, c, d, e.

The first part of Abel's proof requires the notion of the *splitting field* for the polynomial, the smallest field that contains all of its roots. It can be described as the field of all rational functions (quotients of polynomials) in the roots, with coefficients in the same field from which those of the original polynomial were taken. The first part of Abel's proof amounts to showing that, if there is an algebraic formula for the roots, then *the radicals that arise in the course of executing the formula also lie in the splitting field* (see Question 10.2). As Leopold Kronecker (1823–1891) described it, these unavoidable irrationalities in the formula are *natural*, not *extraneous*. We have already seen how this happens in the case of quadratic and cubic equations: The radicals in the coefficients of the polynomial are rational functions when expressed in terms of the roots, and that is merely another way of saying that these radicals lie in the splitting field. We emphasize again, however, that this statement is true only when the original field contains all the necessary roots of unity. What is actually true, is that the radicals are rational functions of the roots of the equation and the roots of unity.

Abel obtained a system of linear equations for these radicals, and the coefficients of the system were a Vandermonde matrix in the mth roots of unity (just as in the case of the quadratic and cubic) and whose right-hand sides were the m roots $x_1, \ldots, x_m$. In other words, the radicals that occur are actually linear functions of the roots, just as Lagrange had pointed out in the particular cases of quadratic and cubic equations. Abel noted that when these equations are solved, one of the results is

$$(4) \qquad R^{1/m} = \frac{1}{m}\left(\alpha_1 x_1 + \alpha_2 x_2 + \cdots + \alpha_m x_m\right),$$

where $\alpha_1, \ldots, \alpha_m$ are the mth roots of unity. (The presence of the roots of unity here shows that this polynomial may not be in the splitting field unless that field also contains the roots of unity.) Since the right-hand side has $m!$ different values as the roots are permuted, it follows that the algebraic formula R must assume $(m-1)!$ different values, that is, R must contain root extractions of orders $m-1, m-2, \ldots, 2$ nested in some order. At that point, the rest of the proof is a matter of counting the number of values certain functions can assume.

3.1. Taking the formula apart.

3.1. Taking the formula apart. It is possible that in the original form (as functions of the coefficients a, b, c, d, e) the function R contains some radical, say an nth root $z = S^{1/n}$, where n is a prime number. If so, the previous argument needs to be repeated to eliminate these radicals. As we strip away the layers of radicals, we continually get expressions that are polynomials in the roots of the original equation under each radical, and the degree of the polynomial indicates the level: An nth root inside an mth root necessarily contains a polynomial in the roots whose degree is a multiple of mn. We saw this in the case of the cubic formula for solving $x^3 - ax^2 + bx - c = 0$, which contains $D_3 = p^3/27 + q^2/4$ under a square root sign under a cube root sign, and this expression (the cubic discriminant) is $-1/108$ times $\left((u-v)(u-w)(v-w)\right)^2$, which is of degree 6.

3.2. The last step in the proof. At the end of his argument, Abel started at the bottom layer of the hypothetical formula for a solution, that is, considering a radical $S^{1/m}$, where S is a *rational* function of the coefficients (and hence a symmetric rational function of the roots) and m is a prime number, necessarily 2, 3, or 5. (Unfortunately, Abel used the letter R instead of S, assuming his readers would know it wasn't the same R he had used earlier.) When expressed as a rational function of the roots, $S^{1/m}$ must assume m different values as the roots are permuted (since all m values of the radical are admissible in the formula). But Ruffini and Cauchy had shown that it is not possible for a rational function of five variables to assume exactly three different values when those variables are permuted. (This result is contained in the exercises to the previous lesson. Even stronger results are known. Cauchy had shown that the number of different values for a rational function of m variables must be either at most 2 or at least m, if m is prime.) Hence there are only the two possibilities $m = 2$ or $m = 5$.

To rule out $m = 5$, Abel noted that $S^{1/m}$ could be expressed in terms of the roots s, t, u, v, w as a multiple of the function $s + \alpha t + \alpha^2 u + \alpha 3v + \alpha^4 w$, where α is a primitive fifth root of unity. But this function assumes 120 values as the roots are permuted, while $S^{1/m}$ assumes only five values. Thus, $m = 5$ is impossible. Therefore $m = 2$, and $S^{1/2}$ is of the form $r + \sqrt{s}$, where s is the discriminant times the square of a symmetric function and r is a symmetric function of the roots. In other words, the hypothetical solution process would have to begin by extracting a square root. (In this connection, see Subsection 2.4 in Lesson 11.)

Then, working back up to the top layer of the formula, Abel argued that a root could be expressed as in Eq. 3, with $m = 5$, so that

$$R^{1/5} = \frac{1}{5}(\alpha_1 x_1 + \alpha_2 x_2 + \alpha_3 x_3 + \alpha_4 x_4 + \alpha_5 x_5) = \left(r + s^{1/2}\right)^{1/5},$$

where $\alpha_1, \ldots, \alpha_5$ are the fifth roots of unity and $x_1, \ldots, x_5$ the five roots of $p(x)$. But, he noted, the middle expression here has $5! = 120$ different values when the roots are permuted, while the right-hand expression has only 10. Thus a contradiction has been reached.

3.3. The verdict on Abel's proof. If some statements in our summary of Abel's proof seem rather vague, do not blame yourself or (even worse) the author. Abel was arguing on a very abstract level about the properties of an algebraic formula for solving the quintic equation. Without a great deal of background in algebra, it is very difficult to be confident that a general algebraic formula must have these properties. The reasons for some of Abel's statements seemed obscure. Indeed, in 1832, only a few years after his publication of this argument, the Prague Scientific Society declared that the question was still open, and offered a prize for a definitive proof. That prize was won by William Rowan Hamilton, who submitted a paper in 1836 that makes for very dense reading.

Once it was accepted that no algebraic formula could be found to express the roots of a general quintic equation, a search was begun for a transcendental formula. Such a formula was discovered by Charles Hermite (1822–1900) in 1858 and by Kronecker in 1861. The formula involves elliptic integrals, whose symmetries had been investigated in great detail by earlier mathematicians, Abel among them. We saw an example of such a transcendental solution in Viète's solution of the irreducible case of the cubic. The formulas given by Hermite and Kronecker, however, are much more complicated.

4. Problems and questions

Problem 10.1. Suppose that $f(x)$ is an algebraic function, that is, there is a non-zero polynomial $p(x, y)$ such that $p(x, f(x)) \equiv 0$. Show that there is some integer n such that $f(x)/x^n \to 0$ and $f(x)x^n \to \infty$ as $x \to \infty$. Then prove that $f(x) = 2^x$ and $f(x) = \sin x$ are not algebraic functions.

Question 10.1. Explain why any algebraic formula $x(a, b, c, d, e)$ for solving the quintic equation would necessarily have to contain, at some point, a cube root. *Hint:* Suppose that $p(x) = 0$ is a cubic equation. You can convert it into a quintic equation by multiplying it by $(x - r)(x - s)$.)

Question 10.2. How do you reconcile the following two facts? (1) Abel showed that the radicals that arise in the course of solving an equation by formula must be in the splitting field of the polynomial. (2) The radicals involved in solving a cubic equation with three real roots (and hence a splitting field consisting of only real numbers) *must* be complex numbers.

5. Further reading

Raymond Ayoub, "Paolo Ruffini's contributions to the quintic," *Archive for History of Exact Sciences*, **23** (1980), pp. 253–277.

W. H. Langdon, "Abel on the quintic equation," in *A Source Book in Mathematics*, David Eugene Smith, ed., Dover, New York, 1959.

Peter Pesic, *Abel's Proof*, MIT Press, Cambridge, MA, 2003.

Michael I. Rosen, "Niels Hendrik Abel and equations of the fifth degree," *American Mathematical Monthly*, **102** (1995), No. 6, pp. 495–505.

Dirk J. Struik, "Euler. The fundamental theorem of algebra," in *A Source Book in Mathematics, 1200–1800*, Dirk J. Struik, ed., Princeton University Press, Princeton, NJ, 1986.

Dirk J. Struik, "Gauss. The fundamental theorem of algebra," in *A Source Book in Mathematics, 1200–1800*, Dirk J. Struik, ed., Princeton University Press, Princeton, NJ, 1986.

Dirk J. Struik, "Girard. The fundamental theorem of algebra," in *A Source Book in Mathematics, 1200–1800*, Dirk J. Struik, ed., Princeton University Press, Princeton, NJ, 1986.

The Breakthrough: Galois Theory

Since formulas exist for solving equations up to degree four, many algebraic numbers can be expressed by applying a finite sequence of rational numbers and root extractions to integers. Algebraic numbers satisfying equations of degree 4 or less with rational coefficients have such an expression.

The converse is not true, however. The absence of a general formula for solving the quintic equation does not imply that an algebraic number whose minimal polynomial is of degree 5 has no such expression. Some of them do, because some particular irreducible equations of degree 5 can be solved by a finite sequence of root extractions and rational operations. An example is the equation $x^5 + 5x^4 + 10x^3 + 10x^2 + 5x - 1 = 0$, which has the solution $x = \sqrt[5]{2} - 1$. The problem is to separate those that have such a solution from those that do not. That problem was taken up by Abel's younger contemporary Évariste Galois when he was in his late teens. (As is well known, Galois died at the age of 20 in a duel. Abel, in contrast, lived to the much riper age of 26 before succumbing to tuberculosis.)

The vagueness in Abel's argument was removed by a closer analysis of the process that leads to a formulaic solution of an equation. Since the formula consists of a series of steps, in each stage of which rational operations and root extractions are applied to expressions obtained at a previous stage, it is not a large step to describe the process by saying that we start with a field containing the coefficients of a given polynomial but none of its roots, then enlarge the field to the smallest one containing one of the roots. If we don't get all the roots at that stage, we continue to enlarge the field until we do. That is, we arrive at what is called the *splitting field* of the polynomial. The quest for an algebraic solution is an attempt to construct this field by enlarging the field one step at a time *by adjoining roots of numbers in the current field*. Notice the essential difference: A root of an *equation* is not necessarily the root of a *number*; it may be a complicated combination of roots of numbers, or not even expressible in terms of roots of numbers. That is the refined problem we are now considering. We are focused on particular equations rather than seeking a general formula to cover all equations.

It is essentially the definition of a field that rational operations applied to its members do not lead to any numbers outside the field. Hence all the enlargements of the field occur when roots are extracted. As Abel pointed out, only roots of prime order need to be extracted, since roots of composite order are obtained by successive extraction of roots of prime order.

Since we are interested primarily in equations of degree 5 or less, and since we know that all square, cube, fourth, and fifth roots of unity can be expressed as algebraic formulas (in fact formulas involving only square roots), we shall assume at first, for the sake of simplicity, that the field $\mathbb{F}$ we start with contains all roots of unity and all the coefficients of the equation we are trying to solve. It necessarily contains all rational numbers, since we are interested only in equations with complex coefficients, and every subfield of the complex numbers contains all the rational numbers.

1. An example of solving an equation by radicals

It will enhance our understanding of the field-extension process to work through it one step at a time for a cubic equation. Since we have an algebraic formula for the roots, we know that they can be reached by adjoining roots of numbers. Let us see how this procedure works using the example $p(x) = 8x^3 + 4x^2 - 4x - 1 = 0$, which is irreducible over the rational numbers, but does have three real roots.

We start with a base field $\mathbb{F}$, which is assumed to contain the rational numbers and all three cube roots of unity but none of the roots. Although we shall not prove this fact at the moment, the polynomial $p(x)$ has no roots in F. We plan to enlarge $\mathbb{F}$ to a larger field $\mathbb{F}'$ containing the three roots. We would like, if possible, for $\mathbb{F}'$ to be minimal, that is, the intersection of all fields containing $\mathbb{F}$ and the three roots of the polynomial. But, as we shall see, if we try to reach it by adjoining roots of *numbers*, we will have to include some extraneous elements. We shall take two routes to this end. The first route is a natural one to try. We shall simply invoke the cubic formula.

The substitution $y = x + 1/6$ produces the standardized cubic equation $8y^3 - (14/3)y - 7/27 = 0$, that is, $y^3 - (7/12)y - 7/216 = 0$. The general formula for the solution is then

$$x = \frac{1}{6}\left(1 + \left(\sqrt[3]{\frac{7}{2} + \frac{21}{2}\sqrt{3}i} + \sqrt[3]{\frac{7}{2} - \frac{21}{2}\sqrt{3}i}\right)\right).$$

Since we assumed that the cube root of unity $\alpha = -1/2 + (\sqrt{3}/2)i$ is in the base field $\mathbb{F}$, the complex number $\sqrt{3}i = 2\alpha + 1$ is in this field, and hence so is the complex number $w = (7 + 21\sqrt{3}i)/2 = 14 + 21\alpha$. In order to evaluate this formula, we require a cube root of this number, which lies outside the real numbers. We also require the complex conjugate of this cube root, which fortunately lies in the field $\mathbb{F}(\sqrt[3]{w})$ obtained by adjoining this cube root to the field $\mathbb{F}$. In fact, since $\bar{w} = |w|^2/w = 343/w$, we find that $\overline{\sqrt[3]{w}} = 7/\sqrt[3]{w}$, which lies in the field $\mathbb{F}(\sqrt[3]{w})$. We can then express one of the roots as

$$x = \frac{1}{6}\left(1 + \sqrt[3]{w} + \frac{7}{\sqrt[3]{w}}\right).$$

Since α belongs to $\mathbb{F}$, we can replace $\sqrt[3]{w}$ by $\alpha\sqrt[3]{w}$ and $\alpha^2\sqrt[3]{w}$ in this formula to get the other two roots. Thus, we can solve the equation by adjoining this one cube root. Notice, however that we have *overshot* the desired splitting field $\mathbb{F}'$, which consists entirely of sums of terms, each of which is a real number multiplied by a root of unity. The enlarged field $\mathbb{F}(\sqrt[3]{w})$ contains the extraneous element $\sqrt[3]{w}$, which is not a sum of this form. This overshooting occurred even though we started with a field containing more elements than were really needed. The equation itself had rational coefficients, so that we could have constructed a field containing all its roots entirely within the real numbers.

Thus our first choice of a number whose root we need to adjoin led us to a field that is larger than it needs to be. But then, the original field was larger than it needs to be, since we included the cube roots of unity in it, and these numbers are not in the splitting field of the polynomial. So should we have expected the enlarged field to contain extraneous elements? Could we have avoided it by not including any in the original field $\mathbb{F}$? No. Even if we hadn't included α at the outset and had begun with the rational numbers $\mathbb{Q}$ as the base field $\mathbb{F}$, we would have had to adjoin α—or, equivalently, $\sqrt{-3}$—in order to compute the number produced by the cubic formula, and so the final field produced would have had to contain α if we used the cubic formula to solve the equation. Following this path and starting with $\mathbb{Q}$, we would have had to adjoin first α or $\sqrt{-3}$, then $\sqrt[3]{w}$, and so there would have been two adjunctions of roots instead of one.

Let us try a second approach. The equation we are now considering can also be solved by adjoining a seventh root of unity that is not itself unity. Since $x^7 - 1 = (x-1)(x^6 + x^5 + x^4 + x^3 + x^2 + x + 1)$, such a root ω satisfies the cyclotomic equation $\omega^6 + \omega^5 + \omega^4 + \omega^3 + \omega^2 + \omega + 1 = 0$. The number $r = (\omega + \omega^6)/2$ is a root of the equation, that is, $p(r) = 0$. All three roots can be produced in this way by choosing different values of ω. But once again, the number ω is a complex number, not in the minimal splitting field, and we have overshot the mark. This example shows that there are subtleties involved in solving equations by adjoining roots of numbers. For one thing, as we just saw, the root whose adjunction solves the equation is not unique.

We can obtain the minimal splitting field by adjoining just one root r of the equation. The other two roots are then easily shown to be $2r^2 - 1$ and $4r^3 - 3r$. But r is not the root of a *number*, so that this way of solving the equation is not a solution by radicals. It appears (and is true) that this equation cannot be solved by radicals *within its minimal splitting field*.

2. Field automorphisms and permutations of roots

To study the general problem of finding a splitting field, we need to see how permuting the roots of a polynomial affects the enlarged field. A typical example occurs when we adjoin a nonreal root r of an equation with real coefficients to a field that originally contained only real numbers. Along with that root, we automatically get its complex conjugate $\bar{r}$, which is also a root. Interchanging the roots $\bar{r}$ and r, or more generally $\bar{z}$ and z for all z in the larger field, preserves the field operations (addition and multiplication) and leaves the elements of the original field invariant. Such a mapping is called a *field automorphism*. In this case, it is a field automorphism leaving the base field invariant, since a real number is its own conjugate.

That situation holds in general. Each permutation of the roots of a polynomial $p(x)$ that have been adjoined to a field $\mathbb{F}$ to produce a larger field $\mathbb{F}'$ results in an automorphism of the field $\mathbb{F}'$ that leaves the field $\mathbb{F}$ invariant. In the example considered in the previous section, the minimal splitting field admits three automorphisms that leave the base field invariant. One is the trivial automorphism that leaves all elements of $\mathbb{F}'$ where they are. Another takes the root r to $2r^2 - 1$, $2r^2 - 1$ to $4r^3 - 3r$, and $4r^3 - 3r$ to r. The third takes r to $4r^3 - 3r$, $4r^3 - 3r$ to $2r^2 - 1$, and $2r^2 - 1$ to r. These are the only possibilities. There is no automorphism that interchanges two of the roots and leaves the third one fixed. (See Problem 11.7, which is a slightly disguised treatment of the equation we have just discussed.)

The larger field $\mathbb{F}(\sqrt[3]{w})$ that we constructed also admits the three automorphisms just listed. Even though it contains complex numbers, this field does not admit complex conjugation as an automorphism, since $\mathbb{F}$ contains α, which must be left fixed. On the other hand, if we had started with just the field of rational numbers as $\mathbb{F}$, omitting the cube roots of unity, then $\mathbb{F}(\sqrt[3]{w})$ *would have* admitted complex conjugation as a field automorphism leaving $\mathbb{F}$ invariant. But complex conjugation leaves all three roots fixed. In that case, we would have had to adjoin α to $\mathbb{F}$ first, before adjoining $\sqrt[3]{w}$ to solve the equation, since the latter is not the root of an element of $\mathbb{F}$. In other words, we would have had to adjoin two radicals instead of one.

Field automorphisms leaving a subfield invariant lie at the heart of Galois theory. The automorphisms of a field form a group G. Those that leave a subfield invariant form a subgroup H of that group. In the context of field extensions, there is a natural one-to-one correspondence between the automorphisms of the splitting field of a polynomial $p(x)$ that leave the base field invariant and the permutations in some subgroup of the group of permutations of the roots of the polynomial being factored. Each automorphism necessarily permutes the roots, since it preserves polynomials with coefficients in $\mathbb{F}$, and in particular preserves $p(x)$. That is, if r is a root and τ is an automorphism, then $\tau(r)$ must also be a root, since $p(\tau(r)) = \tau(p(r)) = \tau(0) = 0$. Conversely, if a permutation of the roots is given, there is at most one automorphism of the larger field that brings about that permutation of the roots in this way.

The set of all automorphisms of the splitting field $\mathbb{F}'$ that leave the base field $\mathbb{F}$ invariant is called the *Galois group* of the equation (or polynomial) over $\mathbb{F}$. Because of the automorphism–permutation correspondence, it is useful to think of the Galois group of a polynomial as a group of permutations of the roots of the polynomial.

We shall now illustrate these concepts using the example of a cubic polynomial $x^3 - ax^2 + bx - c$ with rational or integer coefficients a, b, c and irrational roots u, v, w, taking the rational numbers $\mathbb{Q}$ as a base field.

2.1. Subgroups and cosets. To solve the equation by formula, we first adjoin the square root of the cubic discriminant $d_3 = \sqrt{p^3/27 + q^2/4}$, where $p = b - a^2/3$ and $q = ab/3 - 2a^3/27 - c$. Although the enlarged field $\mathbb{Q}(d_3)$ does not yet contain any of the roots, each of the six permutations of u, v, w generates an automorphism on the enlarged field $\mathbb{Q}(d_3)$ leaving the elements of base field $\mathbb{Q}$ unchanged. Three of them, as noted, generate the trivial automorphism, that is, they also leave the elements of the enlarged field $\mathbb{Q}(d_3)$ unchanged. Those three, the identity permutation that moves nothing and the two cyclic permutations (uvw) and (vuw), form a subgroup of the full symmetric group S_3. This subgroup consists of the even permutations. It is called the *alternating* group on three symbols, and usually denoted A_3 or Z_3. The other three elements of S_3—the three transpositions (vu), (wu), and (vw)—form what is called a *coset* relative to this subgroup. Any two of them differ by an element of the subgroup A_3, since, for example $(vu) = (vuw)(wu)$.

Remark 11.1. The order of cycles that have common elements is not a matter of indifference. Our convention is that $(vuw)(wu)$ means *first* do (vuw), *then* do (wu), that is, starting with the order (u, v, w), put v where u is, u where w is, and w where v is. The result of that is the arrangement (v, w, u). Next do (uw), that is, interchange u and w. The final result is (v, u, w), which is the same as doing (vu) on the original ordering. Thus, as asserted $(vu) = (vuw)(wu)$.

2.2. Normal subgroups and quotient groups. The subgroup A_3 and its single coset $(vu)A_3$ can be made into a group called the *quotient group* of S_3 over A_3. It is in this case a very uncomplicated group consisting of two elements, the set A_3 of even permutations and the set $(vu)A_3$ of odd permutations. The group operation in this case is trivial: even plus even = even = odd plus odd, even plus odd = odd. That is, an odd permutation composed with an even permutation is odd, but the composition of two even or two odd permutations is even. We shall write the composition of two permutations $\tau \circ \sigma$ as $\sigma\tau$. (Note the reversal of the order here. When we regard them as functions, it makes sense to put the operation performed first on the right. But for ease of reading, it makes more sense to proceed from left to right.)

The condition that makes it possible to define a quotient group in general, given a group G and a subgroup H, is that the subgroup H be what is called *normal*. That means the coset to which a composition $\sigma\tau$ belongs depends only on the cosets to which σ and τ belong, not on the particular choice of σ and τ within those cosets. The subgroup A_3 in S_3, mentioned above, is normal, since the product of any two odd permutations is even, the product of any two even permutations is even, and the product of an even permutation with an odd permutation is odd.

In the general case of a normal subgroup, if we choose two other permutations from the same cosets, say $\sigma' = \sigma\lambda$ and $\tau' = \tau\mu$, where λ and μ

belong to H, the product $\sigma'\tau'$ must belong to the same coset as $\sigma\tau$ if H is to be normal. That means $\sigma\lambda\tau\mu = \sigma\tau\nu$ for some $\nu \in H$. Putting it another way, $\lambda\tau\mu = \tau\nu$, or $\tau^{-1}\lambda\tau = \nu\mu^{-1}$, where τ^{-1} represents the inverse of the permutation τ, the one that puts everything back where it was before τ moved it. Since H is a subgroup, $\nu\mu^{-1}$ belongs to H, and so $\tau^{-1}\lambda\tau \in H$.

Since λ was an arbitrary element of H and τ an arbitrary element of G, this last relation says that $\tau^{-1}\lambda\tau$ belongs to H for any $\lambda \in H$ and any $\tau \in G$. Another way of stating the same thing is to say that the left coset τH and the right coset $H\tau$ are the same set. A subgroup of index 2, that is, having only itself and its complement as cosets, is necessarily normal. (The coset λH is H if λ belongs to H and λH is the complement of H otherwise, and the same is true for the right cosets.) An example of a nonnormal subgroup of S_3 is the subgroup K consisting of the identity and a single transposition, say (vu). If we take $\lambda = (vu)$ and $\tau = (vw) = \tau^{-1}$, we find that the left coset $(vw)K$ consists of (vw) and $(vw)(vu) = (wv)(uv) = (wvu)$, while the right coset $K(vw)$ consists of (vw) and $(uv)(wv) = (uvw) = (wuv)$.

2.3. Further analysis of the cubic equation. We now resume our discussion of the field extension process when solving a general cubic equation. After we have adjoined d_3, the enlarged field $\mathbb{Q}(d_3)$ contains the element $r = -q/2 + d_3$. As we saw in Lesson 9, depending on the choice of d_3, this will be either

$$\left(\frac{u + \alpha v + \alpha^2 w}{3}\right)^3$$

if $d_3 = (i/\sqrt{108})(u - v)(u - w)(v - w)$, or

$$\left(\frac{u + \alpha^2 v + \alpha w}{3}\right)^3$$

if $d_3 = (i/\sqrt{108})(v - u)(u - w)(v - w)$. Since the enlarged field contains both d_3 and $-d_3$, it is irrelevant which we choose. Just to keep things simple, we choose the first of these, and take $\sqrt[3]{r} = \frac{1}{3}(u + \alpha v + \alpha^2 w)$.

If we adjoin the element $\sqrt[3]{r} = \frac{1}{3}(u + \alpha v + \alpha^2 w)$ to the field $\mathbb{Q}(d_3)$, to get the larger field $\mathbb{Q}(d_3, \sqrt[3]{r})$, we will be adjoining the cube root of an element in $\mathbb{Q}(d_3)$. We know that the field $\mathbb{Q}(d_3)$, and the element r in particular, is invariant under even permutations of u, v, and w. An odd permutation σ, however, will have the effect of interchanging $r = -q/2 + d_3$ with its conjugate $r^* = -q/2 - d_3$. The group of automorphisms of $\mathbb{Q}(d_3, \sqrt[3]{r})$ that leaves the new base field $\mathbb{Q}(d_3)$ invariant is therefore A_3, the subgroup of even permutations of the roots. This group does *not* leave the newly enlarged field $\mathbb{Q}(d_3, \sqrt[3]{r})$ invariant, since even permutations of the roots can take $\sqrt[3]{r}$ to $\alpha\sqrt[3]{r}$ and $\alpha^2\sqrt[3]{r}$. The enlarged field contains all three roots u, v, w and is *closed* under all permutations of them, but some permutations move the elements of $\mathbb{Q}(d_3, \sqrt[3]{r})$ around.

To sum up, there are six permutations of the roots u, v, w. Each generates an automorphism of the field $\mathbb{Q}(d_3, \sqrt[3]{r})$ that leaves the base field $\mathbb{Q}$ invariant. The even permutations form a subgroup that leaves the larger (intermediate) field $\mathbb{Q}(d_3)$ invariant. There are also permutations of the roots—the three transpositions—that do not leave $\mathbb{Q}(d_3)$ invariant.

The enlarged field $\mathbb{Q}(d_3, \sqrt[3]{r})$ really does contain all three roots. For example, it contains

$$\frac{a}{3} + \frac{a^2 - 3b}{9\sqrt[3]{r}} + \sqrt[3]{r},$$

which, when we substitute $a = u + v + w$, $b = uv + uw + vw$, and $\sqrt[3]{r} = \frac{1}{3}(u + \alpha v + \alpha^2 w)$, becomes

$$\frac{u + v + w}{3} + \frac{u^2 + v^2 + w^2 - uv - uw - vw}{3(u + \alpha v + \alpha^2 w)} + \frac{u + \alpha v + \alpha^2 w}{3} = u.$$

2.4. Why the cubic formula must have the form it does.
Could a solution of the general cubic equation be achieved by adjoining first a cube root of some number in the base field, then a square root? As far as the technique of counting the number of values assumed by a function goes, the order in which root extractions are performed is irrelevant. From that point of view a cubic formula that consists of terms of the form

$$\sqrt{P + Q\sqrt[3]{S} + R(\sqrt[3]{S})^2},$$

where P, Q, R, and S are rational functions of the coefficients a, b, and c, is conceivable. But, as we saw in the previous chapter, Abel was able to rule out this possibility in general, showing that the first root extraction in any general formula must be a square root. More precise proofs of this fact can be found in later papers, such as the paper by Michael Rosen cited at the end of Lesson 10.

By looking at the field extension process, we can see why such a formula is impossible. Before giving the argument, we note that the Tschirnhaus method of solving the cubic might seem to be a contradiction of this principle, since it requires first solving the cubic equation $z^3 = N$, and then solving a quadratic equation in order to get the solution of the original cubic. However, in order to find the N that occurs in the pure equation for z, it is necessary first to find the appropriate substitution z, and that involves solving a quadratic equation. The order of operations when an equation is solved by radicals is reflected in the structure of the Galois group of the equation, as will be explained below.

To see why a formula like the one shown above cannot produce roots of the general cubic equation, let us begin once again with the base field and form some expression S that is a rational function $S(a, b, c)$ of the coefficients and hence a symmetric function of the roots u, v, w. We know that if a cube root of S appears in a general formula for the solution, it must be a rational function of the roots. If we try to find a rational function of the roots, say $f(u, v, w)$, such that $(f(u, v, w))^3 = S(a, b, c)$, we see that permutations of the roots leave S unchanged, since they leave a, b, and c unchanged. It follows that $f(v, u, w)$ must also be a cube root of S. This means that one of the following three possibilities must hold:

$$
\begin{aligned}
f(v, u, w) &= f(u, v, w), \\
f(v, u, w) &= \alpha f(u, v, w), \\
f(v, u, w) &= \alpha^2 f(u, v, w),
\end{aligned}
$$

where α is a primitive cube root of unity. But the second of these implies $f(u, v, w) = \alpha f(v, u, w)$ (interchanging the first two arguments, by assumption, multiplies the function by α), which in turn implies $f(u, v, w) = \alpha^2 f(u, v, w)$, and hence $f(u, v, w) = 0$. Likewise, the third possibility must be ruled out. But that means that $f(u, v, w)$ is invariant under transpositions, and hence invariant under all permutations. In other words, $f(u, v, w)$ is symmetric in the roots, and hence, since it is a rational function, can be expressed in terms of the coefficients a, b, c. Thus $f(u, v, w)$ belongs to the base field $\mathbb{F}$, and we get no enlargement of the field in this way. Thus cannot get a formula for solving a general cubic equation that begins by taking cube roots of rational functions.

2.5. Why the roots of unity are important. When we do not assume that our base field $\mathbb{F}$ contains roots of unity, the field extension process just described becomes more complicated. We can show why, using as an example the equation

$$
x^3 + 6x - 2 = 0.
$$

Because the polynomial $p(x) = x^3 + 6x - 2$ is an increasing function—its derivative $3x^2 + 6$ is always positive—it has a single real zero. Inspection reveals that this root is between 0 and 1. Since the only possible rational roots must be integers that divide 2 (see the Appendix), this polynomial has no roots in the field $\mathbb{Q}$ of rational numbers. Since the equation has one real root and two imaginary roots, we can distinguish the three roots u, v, and w by letting u be the real root, from which it follows that v and w must be $-u/2 \pm yi$ for some positive number y.

We could use the cubic formula to find the roots, but right now we are more interested in the extension process. Let us adjoin the real root u to the base field, getting the larger field $\mathbb{Q}(u)$, which consists of all numbers of the form

$$
r + su + tu^2.
$$

In doing so, we are not "solving by radicals," since the number we adjoined is not a root of a rational number. This field, which consists of real numbers only, does not contain v or w. If there were any nontrivial automorphisms of $\mathbb{Q}(u)$ that leave $\mathbb{Q}$ invariant, the element u would have to map to a root of the equation $x^3 + 6x - 2 = 0$. Since u is the only element of $\mathbb{Q}(u)$ that satisfies this equation, it would have to map to itself. Thus, there are as yet no nontrivial automorphisms leaving $\mathbb{Q}$ invariant. The absence of nontrivial automorphisms of $\mathbb{Q}(u)$ leaving $\mathbb{Q}$ invariant is also reflected in the fact that the subgroup of the complete Galois group of this polynomial leaving $\mathbb{Q}(u)$ invariant is not a normal subgroup, as we shall now see.

There are permutations of u, v, and w that leave the larger field $\mathbb{Q}(u)$ invariant, just as the even permutations left the field $\mathbb{F}(d_3)$ invariant in the general cubic. There are in fact two such permutations, the identity and the transposition (vw), which corresponds to the conjugation automorphism $a + bi \mapsto a - bi$ in the complex numbers and leaves each real number fixed. These form a subgroup H of the full permutation group on three symbols, denoted S_3. As we saw above, *this is not a normal subgroup* of the group S_3, which turns out to be the Galois group of this polynomial. That fact is reflected, as Galois noticed, in the fact that adjoining one root of the equation did not produce a field containing all of its roots. The subgroup H consists of permutations with a fixed point, corresponding to the fact that they must all leave u fixed. But no such subgroup can be a normal subgroup of S_3, since $\sigma^{-1}\tau\sigma$ will not leave u fixed if τ does leave it fixed and σ transposes u with a symbol that τ moves.

In fact, the solution of the equation itself has not been advanced in this way. It is true that over the larger field $\mathbb{Q}(u)$, the polynomial $x^3 + 6x - 2$ factors as $(x - u)((x^2 + ux + (u^2 + 6))$ (since $u^3 + 6u = 2$), and so we could find v and w by solving the quadratic equation

$$x^2 + ux + (6 + u^2) = 0.$$

But we would first need to find u, and the procedure for doing that is to solve the original equation.

Remark 11.2. The cubic formula reveals that $u = \sqrt[3]{4} - \sqrt[3]{2} = \sqrt[3]{2}(\sqrt[3]{2} - 1)$, so that the field extension $\mathbb{Q}(u)$ is contained in the field extension $\mathbb{Q}(\sqrt[3]{2})$. Since both fields are of dimension 3 when regarded as vector spaces over $\mathbb{Q}$, they are in fact, the *same* field.

The other roots v and w are then $\alpha\sqrt[3]{4} - \alpha^2\sqrt[3]{2}$ and $\alpha^2\sqrt[3]{4} - \alpha\sqrt[3]{2}$, where α is a primitive cube root of unity, so that we would get all three roots of the equation in the field extension $\mathbb{F}(u)$ if the base field $\mathbb{F}$ contained all three cube roots of unity. As a consequence, the Galois group of this equation over the field $\mathbb{Q}(\alpha)$ is simply the cyclic group Z_3 of three elements, whose table of operations is the addition table of the field of three elements discussed in Lesson 1.

2.6. The birth of Galois theory. The idea of looking at the way permutations combine in order to study the solution of equations first came to light in France during the late spring of 1832, simultaneously with a paroxysm of political unrest. This unrest was described by Victor Hugo in his great novel *Les misérables*, whose dashing young character Marius Pontmorency has much in common with our present hero Galois, including a radical political commitment, a love gone awry, and a serious gunshot wound sustained during the insurrection of June 1832. Unlike the unlucky Galois, Marius survived and went on to live happily ever after with his love.

The night before the duel that led to his death, Galois wrote to a friend that he had enough material for three papers, one of which was already written. He went on:

> By Propositions II and III of the first paper, one can see a great difference between adjoining one root of an auxiliary equation and adjoining all of its roots.
>
> In both cases, the group of the equation splits into [cosets] such that a fixed permutation takes one into another. But the condition that these [cosets] have the same substitution holds only in the second case [when all the roots are adjoined]. Putting the matter another way, when one group G contains another group H, the group G can be partitioned into [cosets] so that G is the union of the sets H, $H\sigma$, $H\sigma'$,.... It can also be partitioned into [cosets] in such a way that G is the union of the sets H, τH, $\tau' H$,.... Ordinarily, these two partitions are not the same. When they are the same, we call this a *proper partition* [a partition corresponding to a normal subgroup].
>
> It is easy to see that when the group of an equation has no proper partition [no normal subgroups], no matter how one transforms the equation, the groups of the transformed equation will always have the same number of permutations.
>
> In contrast, when the group of an equation admits a proper decomposition, so that it can be partitioned into M [cosets] of N permutations each, the given equation can be solved by means of two equations, one of which will have a group of M permutations and the other a group of N permutations.

We have inserted the modern word *coset* here, where Galois himself used the word *group*. Again, the reason for these assertions is that when the subgroup leaving the elements of the extended field invariant is normal, it means that the extended field contains *all* the roots of the equation of minimal degree (roots of the *minimal polynomial*) satisfied by each of its elements. As Galois said, when the group of the equation contains a (normal) subgroup of N elements having M cosets, then the group essentially splits into the product of the subgroup and the quotient group (consisting of the cosets). Each of these is the Galois group of a simpler equation, corresponding to an algebraic substitution in the original equation.

We have seen this principle at work in the case of the general cubic equation $x^3 - ax^2 + bx - c = (x - a/3)^3 + p(x - a/3) + q = 0$, which splits into two pure equations

$$d_3^2 = \frac{p^3}{27} + \frac{q^2}{4},$$
$$z^3 = r = -\frac{q}{2} - d_3,$$

corresponding to the normal subgroup A_3 consisting of even permutations of the full Galois group S_3, after which we can write the solution of the equation as

$$x = \frac{a}{3} + \frac{p}{3(-q/2 + d_3)}(\sqrt[3]{r})^2 + \sqrt[3]{r}.$$

3. A sketch of Galois theory

What Galois wrote in the passage just quoted sums up Galois theory in general terms, although one may well disagree with the claim that any of this is "easy to see." (Galois got in trouble at a university examination when he told the examiners who asked him for proof of an assertion that it was obvious.) We cannot, of course, prove all these assertions in detail. What we can do is show in general how the theory applies to the solution of equations by radicals.

Let us begin by stating the fundamental facts of Galois theory for the special case of interest to us, in which the base field is the rational numbers $\mathbb{Q}$. We consider a polynomial $p(x)$ that has rational coefficients, but no rational roots, and let $\mathbb{S}$ be its splitting field, that is, the smallest subfield of the field of algebraic (or complex) numbers that contains all the roots of $p(x)$. The *Galois group* of p is the group G of automorphisms of $\mathbb{S}$ leaving each rational number fixed. The following facts are known:

1. The field $\mathbb{S}$ can be regarded as a vector space V over the field $\mathbb{Q}$. As such, it has finite dimension.

2. The Galois group G is finite, and the number of elements in it equals the dimension of the vector space V.

3. Each automorphism in G permutes the roots of $p(x)$, and different automorphisms correspond to different permutations of the roots. Hence the Galois group can be regarded as a group of permutations of the roots.

4. There is a one-to-one correspondence between the subgroups H of the group G and the subfields $\mathbb{K}$ of the field $\mathbb{S}$. Each subfield $\mathbb{K}$ corresponds to the subgroup $H(\mathbb{K})$ consisting of the automorphisms in $\mathbb{G}$ that leave the elements of $\mathbb{K}$ fixed. The subgroup $H(\mathbb{K})$ is the Galois group of $p(x)$ over the field $\mathbb{K}$.

5. If $\mathbb{K}$ is any subfield of $\mathbb{S}$, then $H(\mathbb{K})$ is a normal subgroup of G (so that the quotient group $G/\mathbb{K}$ is defined) if and only if every polynomial that is irreducible over $\mathbb{Q}$ and has one root in $\mathbb{K}$ has all of its roots in $\mathbb{K}$. In that case, $\mathbb{K}$ is naturally called a *normal extension* of $\mathbb{Q}$.

6. If $\mathbb{K}$ is a normal extension of $\mathbb{Q}$, then the group of automorphisms of $\mathbb{K}$ that leave $\mathbb{Q}$ invariant is the quotient group $G/H(\mathbb{K})$.

We note that if $\mathbb{K} = \mathbb{Q}(\sqrt[p]{w})$, where p is prime and $w \in \mathbb{Q}$, then the group $G/H(\mathbb{K})$ is the cyclic group $\mathbb{Z}_p$ of remainders when integers are divided by p, with addition as the group operation. This is very easy to see, since $\mathbb{K}$ consists of numbers of the form $r_0 + r_1\sqrt[p]{w} + r_2(\sqrt[p]{w})^2 + r_n(\sqrt[p]{w})^{p-1}$. An automorphism τ is determined by its effect on $\sqrt[p]{w}$, which must be a pth root of w, and hence a power of $\sqrt[p]{w}$. Thus, τ must permute these powers cyclically, and so the group of automorphisms of $\mathbb{K}$ leaving $\mathbb{Q}$ fixed must be $\tau, \tau^2, \ldots, \tau^{p-1}, \tau^p$. (This last automorphism is the identity, which leaves everything fixed.)

4. Solution by radicals

Let us now start with a base field $\mathbb{F}$ that contains all the roots of unity, along with the rational numbers that it necessarily must contain. Let $p(x)$ be a polynomial with coefficients in $\mathbb{F}$, but no roots in this field. It is possible to extend the field to a larger field in which it has roots (the complex numbers certainly contain all of them). The question is whether one can get from the smaller field to the larger by rational operations and root extractions, in other words, by successive enlargements of the field via extraction of roots of prime order. Let us suppose that it is possible to do so. Starting from a field $\mathbb{K} = \mathbb{F}(\theta_1, \ldots, \theta_n)$ that has been reached at some stage of the enlargement operations, we adjoin the qth root of some element in the field $\mathbb{K}$, say $\theta_{n+1} = \sqrt[q]{\phi}$, where q may be assumed prime. The field $\mathbb{K}(\theta_{n+1}) = \mathbb{F}(\theta_1, \ldots, \theta_n, \theta_{n+1})$ consists of elements $k_0 + k_1 \theta_{n+1} + k_2 \theta_{n+1}^2 + \cdots + k_{q-1} \theta_{n+1}^{q-1}$, where $k_0, \ldots, k_{q-1}$ belong to $\mathbb{K}$. Since the original field $\mathbb{F}$ contains all roots of unity, this enlarged field contains all q of the qth roots of ϕ. It is therefore a normal extension. (Adjoining one root θ_{n+1} of the equation $x^q - \phi = 0$ automatically resulted in the adjunction of all the roots of this equation.) The automorphisms of $\mathbb{K}(\theta_{n+1})$ that leave $\mathbb{K}$ fixed form a group with q elements, and there is "essentially" only one such group. That group is the cyclic group Z_q with q elements, which are the remainders of the integers after division by q, with addition as the group operation. By "essentially," we mean that any two groups with q elements can be matched up element by element in such a way that the product of two elements in one group is always paired with the product of the corresponding two elements in the other group. Such a pairing is called an *isomorphism*.

If this process eventually produces a field in which the polynomial splits, we shall have produced a chain of fields $\mathbb{K} \subset \mathbb{K}_1 \subset \cdots \subset \mathbb{K}_{r-1} \subset \mathbb{K}_r$, in the last of which the polynomial $p(x)$ can be factored completely. Corresponding to them is a set of subgroups of the Galois group $G \supset G_1 \supset \cdots \supset G_{r-1} \supset G_r$, where G is the full Galois group and G_r consists of the identity automorphism alone. G_k consists of the automorphisms of $\mathbb{K}_r$ that leave each element of $\mathbb{K}_k$ fixed. The group of the extension from $\mathbb{K}_k$ to $\mathbb{K}_{k+1}$, which consists of the automorphisms of $\mathbb{K}_{k+1}$ that leave each element of $\mathbb{K}_k$ fixed, can be regarded as the quotient group of G_k modulo G_{k+1}; that is, it consists of the automorphisms of $\mathbb{K}_r$ that leave $\mathbb{K}_k$ fixed (the group G_k) but regards two automorphisms σ and τ as "the same" if $\sigma\tau^{-1}$ leaves $\mathbb{K}_{k+1}$ fixed.

Thus, transferring attention from the field extensions to the Galois group, we see that the equation is solvable by radicals if the Galois group has a decreasing chain of subgroups, each of which is a normal subgroup of its predecessor, and the quotient group of each over its successor is a cyclic group of prime order. A group having this property is called a *solvable* group. It was proved by Walter Feit (1930–2004) and John Thompson (b. 1932) in a famous paper that occupied an entire issue of the 1963 *Pacific Journal of Mathematics* that every group having an odd number of elements is solvable. On the other hand, the alternating group A_5, which consists of the group of even permutations of five symbols, has only two normal subgroups: A_5 itself and the subgroup consisting of the identity permutation alone. It is therefore not solvable, and as a nonobvious consequence, neither is the full symmetric group S_5, which consists of all the permutations of five symbols.

The decreasing chain of subgroups with successive quotients equal to cyclic groups of prime order is in general not unique. However, by a general theorem, the quotient groups in any two such series can be paired up in a way such that each quotient group in one series is paired with a quotient group of the same order in the other series. In the case of the general cubic equation, whose Galois group is the symmetric group S_3, there is only one normal subgroup, namely the subgroup A_3 consisting of the even permutations (the identity and the two 3-cycles). As mentioned above, this group is called the *alternating* group; it has exactly the same group structure as the cyclic group Z_3. Hence for this group, the order of the chain is unique. That is why the cubic formula must have a square root inside a cube root, rather than the other way around.

4.1. Abel's theorem. From our sketch of the way Galois theory works, we can now prove more than Abel set out to prove. Not only is there no general formula for solving a quintic equation, there are even particular quintic equations with rational coefficients whose solutions cannot be expressed as a finite sequence of arithmetic operations and root extractions starting from rational numbers.

In fact, all we need is a polynomial of degree 5 that has three real roots and two complex roots. The polynomial $p(x) = x^5 - 10x + 2$ mentioned in Lesson 1 will do. By Descartes' rule of signs (see the Appendix), it has at most two positive roots and one negative root. Since $p(-2) = -10$, $p(0) = 2$, $p(1) = -7$, and $p(2) = 14$, it does indeed have three real roots s, t, and u and two complex roots v and w, which are conjugates of each other. Now the automorphism $x + iy \mapsto x - iy$ leaves the three real roots fixed and interchanges the two complex roots; that is, it is the simple transposition (vw). Hence the Galois group of this equation over the rational numbers (which do not include all roots of unity, of course) contains at least one simple transposition.

We can extend the base field by adjoining roots one at a time, starting with s. The first enlargement results in a field consisting of elements $p_0 + p_1 s + p_2 s^2 + p_3 s^3 + p_4 s^4$, that is, a field of dimension 5 over the base field. When we get to the splitting field, we will find that its elements can be expressed as linear combinations of $5N$ new elements, and hence the Galois group is a subgroup of S_5 consisting of $5N$ elements. By a general theorem due to Peter Ludwig Mejdell Sylow (1832–1918), any group of $5N$ elements contains an element of order 5. But the only permutations of order 5 in the symmetric group S_5 are the 5-cycles. (See Problem 9.7.) By another general theorem, one that is not difficult to prove, any permutation in S_5 can be written by iterating any given 5-cycle and any given 2-cycle. Hence the Galois group of this equation over the rational numbers is the full symmetric group S_5. But this group is not solvable. It follows, as Galois said, that the solution of this equation cannot be split into the solution of two equations of lower degree. No algebraic substitution will simplify this equation. In particular it is not solvable "by radicals," as an algebraic formula.

Our exploration of some key moments in the history of algebra is now complete. The rest of this final lesson is devoted to some simple examples to make the basic ideas of Galois theory clearer and give some insight into what it can and cannot do.

5. Some simple examples for practice

The extension process illustrated above by the example of the cubic formula is perfectly general. To provide further clarity, we give some elementary examples.

Example 11.1. (*A quadratic extension.*) Let us begin with the equation $x^2 + 1 = 0$. Here the coefficients are rational numbers, indeed integers. Hence our "base" field $\mathbb{F}$ could be any field between the rational numbers and the real numbers. If we adjoin just one root i to this field, we automatically get the second root $-i$ inside the same enlarged field. That enlarged field, which we denote $\mathbb{F}(i)$, consists of all complex numbers $a + bi$ where a and b belong to the original base field $\mathbb{F}$. In particular, if we start with the real numbers $\mathbb{R}$, the enlarged field is all of the complex numbers: $\mathbb{C} = \mathbb{R}(i)$. Adjoining one root of one quadratic equation to the real numbers has led to a field in which every equation whatsoever has a full set of roots!

To illustrate the connection with permutations of roots, we now ask what happens in the enlarged field if we permute the two roots. That is, every number $a + bi$ is interchanged with its *complex conjugate* $a - bi$. As we have now seen several times, the mapping $z = a + bi \mapsto \bar{z} = a - bi$ has two important properties: (1) It preserves the field operations, since $\overline{z + w} = \bar{z} + \bar{w}$ and $\overline{zw} = \bar{z}\bar{w}$; in other words, it is an automorphism of the enlarged field. (2) It leaves the original base field invariant, since $\bar{z} = z$ if z is a real number. Obviously, repeating this permutation restores each number to its original place. To sum up, the *group* consisting of those automorphisms of the enlarged field that leave the base field invariant corresponds to the set of permutations of the two roots, which consists of just two permutations: one that leaves each root where it is (the identity permutation) and one that interchanges the two roots. This group is the Galois group of the equation $x^2 + 1 = 0$. It is the cyclic group of two elements, denoted Z_2.

Example 11.2. (*Another quadratic.*) Consider now the quadratic equation $x^2 + x + 1 = 0$, with some subfield $\mathbb{F}$ of the real numbers as the base field. In the complex numbers, the roots of this equation are the two complex cube roots of unity, $\alpha = -1/2 + (\sqrt{3}/2)i$ and $\bar{\alpha} = -1 - \alpha = \alpha^2$. Obviously, the enlarged field $\mathbb{F}(\alpha)$ automatically contains α^2, and $\alpha = -1 - \alpha^2$ is automatically in the enlarged field $\mathbb{F}(\alpha^2)$. Hence it does not matter which root we adjoin. The polynomial $x^2 + x + 1$ immediately splits in the enlarged field into the product $(x-\alpha)(x+\alpha+1)$. As before $\mathbb{F}(\alpha)$ consists of expressions of the form $r + s\alpha$, where r and s belong to the base field $\mathbb{F}$. Division in this enlarged field is performed using the identity

$$\frac{1}{r + s\alpha} = \frac{r - s}{r^2 - rs + s^2} - \frac{s}{r^2 - rs + s^2}\alpha\,.$$

The denominator in this expression cannot be zero when r and s are real numbers and not both zero, since it equals $(r - s/2)^2 + 3s^2/4$.

How will multiplication be affected if we permute the roots α and $\bar{\alpha}$? Since this mapping is just the complex conjugation $r + s\alpha \mapsto \overline{r + s\alpha}$ that we have already considered, it is indeed an automorphism that leaves the base field invariant. Hence the Galois group of this quadratic equation is also the group of permutations of the two roots. That is, the group is the cyclic group Z_2.

You have no doubt by now guessed that for any quadratic polynomial having coefficients in a field but no roots in that field, the Galois group is the cyclic group Z_2.

Example 11.3. (*A cubic extension.*) Consider the equation $x^3 - 2 = 0$, taking the base field to be the rational numbers $\mathbb{Q}$. This equation has three roots in the complex numbers, one of which—the real number $\sqrt[3]{2}$—we shall denote by p. The other two are the mutually conjugate numbers $p\alpha$ and $p\bar{\alpha}$, where, as in the preceding example, α is a complex cube root of unity.

This time, the enlarged field $\mathbb{Q}(p)$ does not contain any other roots, since the real numbers are a larger field containing the rational numbers and p, but not $p\alpha$ or $p\alpha^2$. The situation is the same one we encountered above when analyzing the equation $x^3 + 6x - 2 = 0$. In fact, $\mathbb{Q}(p)$ consists of all expressions of the form $r + sp + tp^2$, where r, s, and t are rational numbers. It is obvious that when two such expressions are added, the result is an expression of the same form. The product $(r + sp + tp^2)(u + vp + wp^2)$ is also easily seen to be $(ru + 2sw + 2tv) + (rv + su + 2tw)p + (rw + sv + tu)p^2$. It is not quite so obvious how one would divide by a nonzero number. The solution to that problem is provided by the identity

$$\frac{1}{r + sp + tp^2} = \frac{r^2 - st}{r^3 + 2s^3 + 4t^3 - 6rst}$$
$$+ \frac{(2t^2 - rs)p}{r^3 + 2s^3 + 4t^3 - 6rst} + \frac{(s^2 - rt)p^2}{r^3 + 2s^3 + 4t^3 - 6rst}\,.$$

The denominator here cannot be zero if r, s, and t are rational and not all zero. We leave the verification of this fact as an exercise (Problem 11.2).

By the same reasoning, adjoining $p\alpha$ will not bring p or $p\alpha^2$ into the enlarged field, since, for example, the equation $p = r + sp\alpha + tp^2\alpha^2 = (r - tp^2) + (s - tp)p\alpha$ would imply that $\alpha = (p + tp^2 - r)/((s - tp)p)$ is a real number.

Now consider what possible automorphisms there can be in the enlarged field $\mathbb{Q}(p)$ that leave the base field $\mathbb{Q}$ invariant. Since $p^3 = 2$, any such automorphism $z \mapsto z^*$, would have to be such that $(p^*)^3 = (p^3)^* = 2^* = 2$. Hence p^* would have to be a cube root of 2 in $\mathbb{Q}(p)$. But p is the *only* cube root of 2 in this minimally enlarged field. Therefore the automorphism would have to be $(r + sp + tp^2)^* = r + sp^* + t(p^*)^2 = r + sp + tp^2$. In other words, this group of automorphisms is the trivial group consisting of the identity element alone.

Now, if we enlarge the new field by adjoining $p\alpha$ as a second root, we automatically get not only the third root $p\alpha^2 = -p\alpha - p$ in the enlarged field $\mathbb{Q}(p, p\alpha)$, but also α, since $\alpha = p\alpha/p$. Conversely, adjoining α to $\mathbb{Q}(p)$ brings in $p\alpha$ and $p\alpha^2$, so that the newly enlarged field is exactly what we would get by adjoining α to the field $\mathbb{Q}(p)$. That is convenient for us since, by what we showed above, it means we can express every element of the newly enlarged field in the form

$$(r + sp + tp^2) + (u + vp + wp^2)\alpha .$$

Thus, if we first solve the auxiliary equation $x^2 + x + 1 = 0$, the splitting field of the original equation becomes easier to characterize. At this point, it is no surprise that starting with a base field containing all cube roots of unity simplifies both the extension process and the Galois group.

Now that we have a field large enough to split the polynomial $z^3 - 2 = (z-p)(z-p\alpha)(z-p\bar{\alpha})$, but not any larger than it needs to be for this purpose, we can ask again, what are the automorphisms of this field that leave the rational numbers invariant? Since every element is written in terms of p and α, we are merely asking where these elements must map. From what was shown above, α must map either to itself or to $\bar{\alpha} = -1 - \alpha$, and the latter mapping is a genuine possibility, since it leaves not only $\mathbb{Q}$, but also $\mathbb{Q}(p)$ invariant. Are there any others? We know that p would have to map to another cube root of 2; that is, the map would have to be $p^* = p\alpha$ or $p^* = p\bar{\alpha} = \overline{p\alpha}$. Is the mapping $p^* = p\alpha$ a genuine possibility? If we keep α fixed, this entails that $(p\alpha)^* = p\alpha^2 = p\bar{\alpha}$ and $(p\bar{\alpha})^* = p\alpha\bar{\alpha} = p$. In other words, it gives a cyclic permutation of the roots: $p \mapsto p\alpha \mapsto p\bar{\alpha} \mapsto p$. Similarly, keeping α fixed and mapping p to $p\bar{\alpha}$ would give the cyclic permutation $p \mapsto p\bar{\alpha} \mapsto p\alpha \mapsto p$. Since the mapping $\alpha \mapsto \bar{\alpha}$ permutes $p\alpha$ and $p\bar{\alpha}$, combining this last automorphism with the first two provides four different automorphisms. Applying conjugation alone and leaving all roots fixed gives us a total of six automorphisms, and shows that the Galois group of this equation over the rational numbers consists of the whole group S_3 of six permutations of the roots. (Each permutation determines an automorphism and vice versa.)

Notice that we got two different sequences of field extensions leading from the base field $\mathbb{Q}$ to the splitting field $\mathbb{Q}(\sqrt[3]{2}, \alpha) = \mathbb{Q}(\alpha, \sqrt[3]{2})$. These were $\mathbb{Q} \subset \mathbb{Q}(\sqrt[3]{2}) \subset \mathbb{Q}(\sqrt[3]{2}, \alpha)$ and $\mathbb{Q} \subset \mathbb{Q}(\alpha) \subset \mathbb{Q}(\alpha, \sqrt[3]{2})$. Although the full Galois group of automorphisms of the splitting field leaving the base field invariant is the same at the end, namely S_3, the behavior of the intermediate field is different in the two cases, and this behavior is reflected in the structure of the Galois group.

In the first case, there are no nontrivial automorphisms of $\mathbb{Q}(\sqrt[3]{2})$ that leave $\mathbb{Q}$ invariant. If we adjoin $\sqrt[3]{2}$ first, we do not get all the roots as a result, but instead are left with a quadratic equation that we need to solve, namely $x^2 + \sqrt[3]{2}x + \sqrt[3]{4} = 0$. In other words, this is not a *normal* extension. The quadratic formula reveals that the roots of this equation are $\sqrt[3]{2}\alpha$ and $\sqrt[3]{2}\alpha^2$, so that we need to adjoin α in order to split the polynomial. The absence of nontrivial automorphisms at the initial extension is reflected in the final extension, in the subgroup of automorphisms of $\mathbb{Q}(\sqrt[3]{2}, \alpha)$ that leave $\mathbb{Q}(\sqrt[3]{2})$ invariant. This subgroup consists of the identity and complex conjugation; it interchanges the two complex roots and leaves the real root fixed. Hence it cannot be a normal subgroup of S_3. In other words, extending the field in this order produced the Galois group, but no normal subgroup of it.

In the second case, the field $\mathbb{Q}(\alpha)$ does have a nontrivial automorphism, complex conjugation, that leaves $\mathbb{Q}$ invariant. Again this fact is reflected in the final extension, in the subgroup of the Galois group that leaves $\mathbb{Q}(\alpha)$ invariant. This subgroup consists of three elements: (1) the identity automorphism; (2) the cycle $p \mapsto \alpha p \mapsto \alpha^2 p \mapsto p$; (2) the cycle $p \mapsto \alpha^2 p \mapsto \alpha p \mapsto p$. This is the subgroup A_3 consisting of even permutations of the roots, and it is the Galois group of the equation $x^3 - 2 = 0$ over the field $\mathbb{Q}(\alpha)$. In this case the group of automorphisms from the first extension is the quotient of the full Galois group S_3 over the group of automorphisms from the second extension. The subgroup A_3 is normal, and its quotient group is the two-element group corresponding to the conjugation operation $\alpha \mapsto \bar{\alpha} = \alpha^2$. The latter is an automorphism of $\mathbb{Q}(\alpha)$ that leaves $\mathbb{Q}$ invariant.

Again, this group structure reflects the solution process. If we adjoin α, then $\sqrt[3]{2}$, in each case we progress from an equation having no roots in the preceding field to one having all its roots in the next field. In the second case, the Galois group is the quotient group of the group of automorphisms leaving $\mathbb{Q}$ invariant over the normal subgroup of automorphisms leaving $\mathbb{Q}(\alpha)$ invariant. In the first case, there is no such decomposition into smaller groups.

Instead of α, we could have joined $\sqrt{-3}$ (which is $2\alpha + 1$) first. In that case, we would have obtained a second intermediate field $\mathbb{Q}(\sqrt{-3})$ that is a normal extension, and the subgroups of the Galois group in this extension and the previous one would pair off in such a way that the quotient groups were isomorphic.

Example 11.4. (*A quartic polynomial.*)   The three simple examples we gave above may lead you to conclude that the Galois group of every irreducible polynomial is the group of all permutations of its roots. To see why that conjecture is wrong, consider the equation $x^4 + 1 = 0$, with $\mathbb{Q}$ again as the base field. Any root r of this equation will satisfy $(r^2)^2 = -1$, and therefore $r^2 = \pm i$. Hence the enlarged field $\mathbb{Q}(r)$ will contain all complex numbers of the form $a + bi$, where a and b are rational numbers. Since the four roots of the equation in the complex numbers are $r_1 = (1 + i)/\sqrt{2}$, $r_2 = (1 - i)/\sqrt{2}$, $r_3 = (-1 + i)/\sqrt{2}$, and $r_4 = (-1 - i)/\sqrt{2}$, we see that the enlarged field will contain $\sqrt{2}$ as well, no matter which root is adjoined. For example, $\sqrt{2} = (1+i)/r_1$. Hence the polynomial splits completely whenever any one of its roots is adjoined, and any element of it can be written in the form

$$r + is + (t + iu)\sqrt{2},$$

where r, s, t, and u are rational numbers.

Now we ask which automorphisms of the enlarged field leave the base field invariant. As always, conjugation is a possibility, in this case actually *two* possibilities. Obviously since -1 and 2 are invariant, there are only four possibilities: $i \mapsto \pm i$ and $\sqrt{2} \mapsto \pm\sqrt{2}$. The choice $i \mapsto -i$, $\sqrt{2} \mapsto \sqrt{2}$ interchanges r_1 with r_2 and r_3 with r_4. The choice $i \mapsto i$, $\sqrt{2} \mapsto -\sqrt{2}$ interchanges r_1 with r_4 and r_2 with r_3. Finally, the choice $i \mapsto -i$, $\sqrt{2} \mapsto -\sqrt{2}$ interchanges r_1 with r_3 and r_2 with r_4. Since each of these is an interchange of disjoint pairs of roots, each, when repeated, restores the roots to their original place. The resulting Galois group of four permutations is a famous group, known as the *Klein four-group*. It is the same group shown as the field of four elements with addition as the group operation (see Problem 6.3).

Remark 11.3. We could have achieved this extension in two steps if we had wanted, by first adjoining $\sqrt{2}$ to the rational numbers (solving the auxiliary equation $x^2 - 2 = 0$). When that is done, the polynomial $x^4 + 1$ is no longer irreducible, since $x^4 + 1 = (x^2 + \sqrt{2}x + 1)(x^2 - \sqrt{2}x + 1)$. Either one of the quadratic factors would have produced the field we finally arrived at. In this way, we would have had a quadratic extension of a quadratic extension, and it would be obvious that we are dealing with a group of two permutations extended by a group of two permutations, that is, a group G having a normal subgroup of two elements, for which the quotient group also has two elements.

Example 11.5. Like many mathematical theories, Galois theory has much in common with politicians: It offers generous amounts of information that you either knew already or never wanted to know, but tends to be very confusing and evasive when answering the question you did ask. A glance through any book on Galois theory will reveal that, in their examples, authors often seem to start their construction of the Galois group *already knowing* the roots of a given polynomial (as we have done in the preceding examples). Theoretically, it ought to be possible to start from the coefficients, rather than the roots. After all, we know that the coefficients determine the roots and the roots determine the Galois group. In the schema that goes

$$(\text{coefficients}) \longrightarrow (\text{roots}) \longrightarrow (\text{Galois group}),$$

we really would like to find some way of proceeding directly from the first to the last, without having to go through the middle. Surely the purpose of the Galois group is to find a path to the solution of an equation, is it not?

Well, actually, the famous classical applications of Galois theory have been in the construction of impossibility proofs. It put the capstone on the proof that there is no formula for solving the general quintic, as we have seen, and it shows that there are no straightedge-and-compass constructions that will trisect a 60° angle or produce the side of a cube twice as large as a given cube, or produce the side of a square equal in area to a circle of given radius. So, is Galois theory a mathematical theory rich in negative results? If so, its applicability would seem to be limited.

For the moment, let us justify the time and effort spent creating Galois theory by simply admiring the human minds that have uncovered such deep algebraic secrets. Even if no way of computing the Galois group directly from the coefficients had ever been discovered, Galois theory would still be a splendid example of the power of human symbol-making capacity. In the equation $x^5 - 10x + 2 = 0$ that we considered above, we don't know what the five roots s, t, u, v, w are, but we can invent symbols to stand for them, and we can reason about them using only the properties that we do know, to conclude finally that they cannot be expressed as finite algebraic formulas involving only rational numbers. In doing so, we are following the technique of Pappus known as *analysis* and mentioned in Lesson 1.

The fact is, however, that a Galois group can be very difficult to calculate unless you know some elements of the splitting field of the polynomial or some sophisticated theorems about finite groups and fields. Algorithms do exist for finding the Galois group starting from the coefficients, or at least narrowing the range of possibilities for it. One such algorithm is encoded in the computer algebra program *Maple*. A very powerful tool was provided by Nikolai Grigorevich Chebotarev (1894–1947), whose name is pronounced "ChebotarYAWF." The Chebotarev Density Theorem provides some principles that enable the range of possible Galois groups to be restricted, in some cases leaving only one possibility.

With the limited techniques we are assuming here, the problem seems circular. If you can get detailed information about the splitting field of the polynomial, you can find the Galois group. But if you have that information, you can also, in all probability, solve the equation without knowing the Galois group. You may well wonder what the Galois group contributes. The applications of this theory depend on knowing some facts about finite groups and fields, and even with those facts at your disposal, you may have to work fairly hard in order to calculate a nontrivial Galois group.

On the positive side, if you do happen to know the Galois group, you can get some useful information about its roots and how to find them. Two problems must be solved when applying Galois theory to solve an equation: (1) Compute the Galois group that the problem leads to; and (2) interpret its structure in terms of the problem at hand in order to get the solution. Problem 1, as already mentioned, is quite difficult in general. Problem 2 may also require more than trivial considerations and perhaps some "heavy artillery," in the form of a computer, to soften it up a bit.

For example, theoretically, the decomposition of a solvable Galois group into a sequence of subgroups with simple quotient groups as factors, tells you to adjoin roots of specified order to the base field in a certain order. The question is: *Which* radicals? We saw this problem in the previous chapter, when we investigated ways of solving the equation $8x^3 + 4x^2 - 4x - 1 = 0$. As another example, although the general fifth-degree polynomial is not solvable by radicals, some particular fifth-degree polynomials are, namely those whose Galois groups are solvable subgroups of S_5. Research into this problem continues, even today, using computer algebra programs like *Mathematica* (see the paper by Dummit in the literature at the end of this lesson) and *Maple* (see the paper by Lazard, also in the literature at the end of this lesson). Since we don't have space to develop a sophisticated theory of fields and finite groups, we shall confine ourselves to a "barehanded" technique of getting detailed information on a field extension.

To get such information about an extension $\mathbb{K}(\theta)$ of a field $\mathbb{K}$ containing the coefficients of a polynomial $p(x)$, you need to "coax" the equation into giving up some information. We shall illustrate with an example, but we warn the reader in advance that the information we are about to produce concerning $\mathbb{K}(\theta)$ is obtained by some rather artificial combinatorial work that may be better described as heavy arm-twisting rather than mere coaxing.

The equation we choose is

$$x^6 + 9x^4 - 4x^3 + 27x^2 + 36x + 31 = 0.$$

The base field here is the rational numbers $\mathbb{Q}$. This polynomial is irreducible over this field, since any solution of it would have to be an integer that divides 31, that is ± 1 or ± 31 (see the Appendix), and it is easy to see that none of these numbers is a root. Even more can be said: Since $27x^2 > 4x^3$ for $0 \leq x \leq 1$ and $9x^4 > 4x^3$ for $x > 1$, it follows that $p(x) \geq 31$ for $x \geq 0$, so that there are not even any positive real roots. As for negative roots, consider what happens when x is replaced by $-x$. All terms now have positive coefficients except x itself. But since $36^2 - 4 \cdot 27 \cdot 31 = -2052$, it follows that $27x^2 - 36x + 31 > 0$ for all real x, and therefore there are no negative real roots, either. Hence the roots of this equation consist of three conjugate pairs of complex numbers $u_1 \pm iv_1$, $u_2 \pm iv_2$, and $u_3 \pm iv_3$.

We now adjoin a root θ of this equation to the rational numbers to produce the larger field $\mathbb{Q}(\theta)$ consisting of all expressions of the form

$$r_0 + r_1\theta + r_2\theta^2 + r_3\theta^3 + r_4\theta^4 + r_5\theta^5.$$

Alternatively, $\mathbb{Q}(\theta)$ can be described as all rational functions of θ

$$R(\theta) = \frac{q(\theta)}{r(\theta)},$$

where $q(x)$ and $r(x)$ are polynomials with rational coefficients.

Neither of these descriptions gives any real insight into the concrete nature of the field $\mathbb{Q}(\theta)$. It remains a bleak, fog-covered landscape in which we can distinguish nothing.

However, by manipulating the equation, rewriting it, we can—after some thought and experiment—produce the following equivalent forms:

$$\begin{aligned}
-27\theta^4 + 54\theta^2 - 27 &= \theta^6 - 18\theta^4 - 4\theta^3 + 81\theta^2 + 36\theta + 4\,, \\
-27(\theta^2 - 1)^2 &= (\theta^3 - 9\theta - 2)^2\,, \\
-3 &= \left(\frac{\theta^3 - 9\theta - 2}{3(\theta^2 - 1)}\right)^2\,.
\end{aligned}$$

This last equation shows that $\mathbb{Q}(\theta)$ contains a number a such that $a^2 = -3$. Since it also contains $-a$, we now know a few non-trivial complex numbers in the field $\mathbb{Q}(\theta)$, namely $\pm\sqrt{3}i$, and consequently also the primitive cube roots of 1, which are $\alpha = -1/2 + (\sqrt{3}/2)i$ and $\alpha^2 = -1/2 - (\sqrt{3}/2)i$.

Can we perform this trick again? It turns out that we can, if we make use of the relations

$$\begin{aligned}
\theta^7 &= -9\theta^5 + 4\theta^4 - 27\theta^3 - 36\theta^2 - 31\theta\,, \\
\theta^8 &= -9\theta^6 + 4\theta^5 - 27\theta^4 - 36\theta^3 - 31\theta^2\,, \\
\theta^9 &= -9\theta^7 + 4\theta^6 - 27\theta^5 - 36\theta^4 - 31\theta^3\,, \\
&= 4\theta^6 + 54\theta^5 - 72\theta^4 + 212\theta^3 + 324\theta^2 + 279\theta\,.
\end{aligned}$$

Using these relations, we can show that

$$27(\theta^2 - 1)^3 = 4(\theta^3 + 3\theta + 1)^3\,,$$

that is,

$$2 = \left[\frac{2}{3}\left(\frac{\theta^3 + 3\theta + 1}{\theta^2 - 1}\right)\right]^3\,.$$

It now follows that the field $\mathbb{Q}(\theta)$ contains an element b such that $b^3 = 2$. Then it must also contain $b\alpha$ and $b\alpha^2$, that is, all three complex cube roots of 2, and in particular the real cube root $\sqrt[3]{2}$.

The fog has now lifted completely from the field $\mathbb{Q}(\theta)$. We now know that it consists of all elements of the form

$$r + s\sqrt[3]{2} + t\sqrt[3]{4} + (u\sqrt{3} + v\sqrt[3]{2}\sqrt{3} + w\sqrt[3]{4}\sqrt{3})i\,,$$

where r, s, t, u, v, and w are rational numbers. As you can easily see, $\mathbb{Q}(\theta)$ is exactly the same extension field that we obtained above for the equation $x^3 - 2 = 0$, where we wrote $p = \sqrt[3]{2}$.

We did not make use of any special properties of θ, other than its being a root of the equation $p(x) = 0$. It follows that any one of the six roots θ would have generated this same field extension. Hence, when we adjoin just one of the six roots, we automatically get a field that contains all six roots. In other words $\mathbb{Q}(\theta)$ is the splitting field of this polynomial. It follows that the Galois group of this equation is S_3.

At this point, the equation itself is practically solved, since we know the form that all six roots must have, and that they must occur in conjugate pairs. The actual solution of this sextic equation is left as an exercise (Problem 11.4). It turns out that, for once, we can solve by radicals (adjoining $\sqrt[3]{2}$ and $\sqrt{-3}$) without overshooting the splitting field.

6. The story of polynomial algebra: a recap

Galois theory, which comes at the end of the story we are telling, is actually the opening of an entirely new era in mathematics. We have only hinted at the challenging problems it presented and continues to present. Having reached the end of the portion of the story we wish to tell, let us cast one quick look back at the milestones that mark this intellectual journey of the human race.

1. The first realization that a number could be described without being explicitly named occurs in ancient texts, which show how to use arithmetic to solve problems that we can interpret as linear equations or systems of linear equations.

2. Some of these ancient texts also solve problems that we interpret as quadratic equations. The key to doing so is the identity $(u+v)^2 \equiv (u-v)^2 + 4uv$, which allows each of the sum and difference of two unknown numbers to be computed from the other if their product is also known.

3. Many problems that were studied by the ancient Greeks entirely within the scope of plane and solid geometry, without algebra, were found by Muslim mathematicians to be related to cubic equations. Their work was built upon by the Renaissance Italian mathematicians, producing a solution of the cubic equation based on the identity $(u-v)^3 + 3uv(u-v) + (v^3 - u^3) \equiv 0$. This achievement marked the limit of applicability of such combinatorial methods alone in solving equations.

4. Finding the algebraic substitution needed to solve the quartic equation by formula required the solution of an auxiliary cubic equation called the *resolvent*, and opened up a new approach to the general problem of finding roots of polynomials, the search for a resolvent.

5. Two centuries of searching for a resolvent to solve the quintic equation finally brought the realization that no such resolvent could exist. This realization came about as the result of the study of permutations of the roots of a polynomial, and eventually led beyond the mere non-existence of a general algebraic-formula solution to prove the non-existence of finite algebraic expressions for even some particular equations.

7. Problems and questions

Problem 11.1. Prove that the Diophantine equation

$$u^3 + 2v^3 + 4w^3 = 6uvw$$

has only $(0, 0, 0)$ as a solution. (Notice that if a prime number p divides two of the three integers u, v, w, then it must divide all three of them, and p^3 can be divided out of the equation. By doing this with as many primes as possible, we can produce a second solution in which any two of u, v, and w have no common prime factors. Use the same letters to denote this second solution. Observe that u must be even and hence v and w both odd. Assume $u = 2u'$, and derive a contradiction.)

Problem 11.2. Prove that the only rational numbers r, s, t satisfying

$$r^3 + 2s^3 + 4t^3 = 6rst$$

are $r = s = t = 0$. *Hint:* Suppose that $r = a/a'$, $s = b/b'$, and $t = c/c'$ satisfy this relation. Consider the integers $u = ab'c'$, $v = a'bc'$, $w = a'b'c$.

Problem 11.3. Prove that the splitting fields of the equations $x^3 - 2 = 0$ and $x^6 + 9x^4 - 4x^3 + 27x^2 + 36x + 31$ over $\mathbb{Q}$ are the same, and hence the Galois groups are the same, namely the full symmetric group S_3.

Problem 11.4. Prove that if $a^2 = -3$ and $b^3 = 2$, then $a + b$ is a solution of the equation

$$x^6 + 9x^4 - 4x^3 + 27x^2 + 36x + 31 = 0.$$

Problem 11.5. Compute the Galois group (over $\mathbb{Q}$) of the equation

$$x^6 - 6x^4 - 4x^3 + 12x^2 - 24x - 4 = 0,$$

whose solutions we already know from Lesson 1. *Hint:* Any field extension $\mathbb{Q}(\theta)$ that contains the two complex cube roots of 2 has to contain the six-dimensional splitting field of the polynomial $x^3 - 2$. Show that this field must be extended by adjoining $\sqrt{2}$ in order to split the present polynomial.

Problem 11.6. Prove that the Galois group of any irreducible polynomial of degree higher than 1 over the real numbers is Z_2. Thus, Galois theory over the field of real numbers is very trivial.

Problem 11.7. As we saw in our examples of the equations $x^3 + 6x - 2 = 0$ and $x^3 - 2 = 0$ (Example 11.3), the splitting field of a cubic equation may be six-dimensional when regarded as a vector space over the base field. In these two cases, where we started without the cube roots of unity in the base field and the equation has only one real root, adjoining that root results in a field contained in the real numbers. It is then necessary to adjoin a second root in order to get the full splitting field. If the base field is enlarged to include cube roots of unity, a single root adjunction suffices and produces a three-dimensional extension field.

We also saw, in our example of the polynomial $p(x) = 8x^3 + 4x^2 - 4x - 1$, that even when the splitting field is three-dimensional, the field produced via a solution by radicals may be larger than the splitting field. It is in general necessary to include complex numbers in the final field enlargement in order to solve a cubic equation having three distinct real roots, what we called the "irreducible case." Consider the polynomial $p(x) = x^3 + x^2 - 2x - 1 = 0$. By examining its values at $x = -2$, $x = -1$, $x = 0$, $x = 1$, and $x = 2$, you can see that this polynomial must have a root between -2 and -1, another between -1 and 0, and a third between 1 and 2. Let r denote the root that lies between 1 and 2. Starting from the equation $r^3 = -r^2 + 2r + 1$, derive successively the equations $r^4 = 3r^2 - r - 1$, $r^5 = -4r^2 + 5r + 3$, and $r^6 = 9r^2 - 5r - 4$. Then use these equations to show that $s = r^2 - 2$ is also a solution of the equation. Since the sum of the roots must be -1, it follows that the third root is $t = -1 - r - s = 1 - r - r^2$. Hence the field $\mathbb{Q}(r)$ is the splitting field of this equation, and is three-dimensional as a vector space over $\mathbb{Q}$.

Show that any automorphism of $\mathbb{Q}(r)$ that leaves $\mathbb{Q}$ fixed must map s to t and t to r if it maps r to s. Similarly, it must map t to s and s to r if it maps r to t. It therefore follows that the Galois group of this equation is Z_3.

Solve this equation by the Cardano method, using the substitution $x = y - \frac{1}{3}$. What roots of complex numbers need to be adjoined in order to compute this solution? Why are these roots *not* in the field $\mathbb{Q}(r)$? This problem shows that when solving an equation by radicals, it is sometimes necessary to "overshoot" the minimal splitting field of the equation. It also shows that even a simple Galois group, which guarantees that the equation can be solved by radicals, does not give any obvious clue as to *which* radicals need to be adjoined to solve the equation.

Finally, solve the equation by Viète's method to get a solution

$$x = -\frac{1}{3} + y = -\frac{1}{3} + \frac{2\sqrt{7}}{3} \cos\left(\frac{\beta}{3}\right),$$

where

$$\cos(\beta) = \frac{1}{2\sqrt{7}}.$$

Problem 11.8. Galois theory over a finite field develops complications not present in the case of a subfield of the complex numbers. The quadratic formula shows that in any field where $1 + 1 \neq 0$, every quadratic equation can be solved by adjoining the square root of a suitable element (depending on the equation). Show that this is not the case in the field of two elements, where it is necessary to adjoin two cube roots of unity in order to solve the equation $x^2 + x + 1 = 0$, and that these two cube roots cannot be expressed in the form $a + b\sqrt{c}$, where a, b, and c belong to the field of two elements.

Question 11.1. Why can't a 60° angle be trisected using straightedge and compass? Give a heuristic argument along the following lines: If you could do so, you could locate the point $(\cos 20°, \sin 20°)$, and hence by projection, the point $(\cos 20°, 0)$. That is, you could construct the real number $\cos 20°$. Show that the polynomial of lowest degree with rational coefficients that this number satisfies is $8x^3 - 6x + 1 = 0$. But the minimal polynomial for a Euclidean-constructible number (a length that can be constructed with straightedge and compass) cannot be of degree 3, since these numbers are obtained as the solutions of quadratic equations whose coefficients satisfy quadratic equations whose coefficients... satisfy quadratic equations with rational coefficients. (Each new point that can be located is the intersection of a circle $(x - h)^2 + (y - k)^2 = r^2$ and a line $ax + by = c$, where a, b, c, h, k, and r have been constructed previously.)

8. Further reading

David S. Dummit, "Solving solvable quintics," *Mathematics of Computation*, **57** (1991), No. 195, pp. 387–401.

Harold M. Edwards, *Galois Theory*, Springer-Verlag, New York, 1984.

R. Bruce King, *Beyond the Quartic Equation*, Birkhäuser, Boston, 1996.

Felix Klein, *The Icosahedron and the Solution of Equations of the Fifth Degree*, Dover, New York, 1956.

Daniel Lazard, "Solving quintics by radicals," in *The Legacy of Niels Henrik Abel*, Springer, Berlin 2004.

Helena M. Pycior, "The philosophy of algebra," in *Companion Encyclopedia of the History and Philosophy of the Mathematical Sciences*, Vol. 1, I. Grattan-Guinness, ed., Routledge, London, 1994.

L. Toti Rigatelli, "The theory of equations from Cardano to Galois, 1540–1830," in *Companion Encyclopedia of the History and Philosophy of the Mathematical Sciences*, Vol. 1, I. Grattan-Guinness, ed., Routledge, London, 1994.

Ian Stewart, *Galois Theory*, Chapman and Hall, London, 1973.

Jean-Pierre Tignol, *Galois' Theory of Algebraic Equations*, World Scientific, Singapore, 2004.

Louis Weisner, "Galois on groups and equations and elliptic integrals," in *A Source Book in Mathematics*, David Eugene Smith, ed., Dover, New York, 1959.

Epilogue: Modern Algebra

The closing off of the search for general algebraic formulas to solve higher-degree equations in the midnineteenth century did not mean that algebra was "over and done with." The solution of this problem was achieved only by passing to a higher level of abstraction than mathematicians had been accustomed to. Instead of merely counting the number of values that a function could assume when its variables were permuted, as Cauchy had done, Galois found it necessary to look at the permutations themselves and distinguish between proper and improper decompositions of the group, or, as we now say, between normal and nonnormal subgroups. This shift in emphasis marked the beginning of a gigantic paradigm shift in mathematics, not only in algebra but also in geometry, number theory, and analysis. All areas marched in the direction of increasing abstraction.

Although algebra was the first area to undergo this increase in abstraction, no area of mathematics was left behind. Where earlier geometers had studied surfaces and curves in Euclidean space, Riemann introduced the notion of an n-fold extended quantity, what we now call a *manifold*. Both geometry and the need to study the convergence of trigonometric series led Georg Cantor (1845–1918) to create an abstract theory of point sets during the 1880s, which then formed the basis for new theories of integration. Number theorists began to move beyond the study of Diophantine equations and the divisibility properties of positive integers, taking up the study of more abstract entities such as the *Gaussian integers* (complex numbers of the form $m + ni$, where m and n are integers) or even more abstract "ideal numbers" introduced by Ernst Eduard Kummer (1810–1893).

Algebra was the leader in all this abstraction and became the indispensable underpinning of many other areas of mathematics, such as topology and analysis, as they increased in abstraction. We shall close out our discussion by informally describing some of the more common and useful specimens of these abstract structures. What follows is a whirlwind tour through the zoological park of modern algebra.

1. Groups

We have made extensive use of permutation groups in the preceding lessons, so that the reader may already have some intuitive picture of what a group is. The word itself, as we saw, is due to Galois.

Modern algebra books present groups axiomatically. This axiomatization was a long time in coming, from the first introduction of the notion of a group of permutations of the roots of an equation by Galois in 1830. Four decades later, when Sophus Lie (1842–1899) and Felix Klein (1849–1925) wanted to create an analog of Galois theory for differential equations, they still thought of groups as sets of permutations, one-to-one mappings that obeyed a cancellation law, that is, $ab = ac \implies b = c$. Such a law automatically holds when a, b, and c are one-to-one mappings and ab means the composition of the two mappings. The abstract notion of a group was well established by the early twentieth century, but Klein himself did not think highly of it. He claimed that, while the abstract formulation was useful in writing elegant proofs, it did nothing to help the mathematician discover new ideas and methods.

Leaving aside the question whether Klein was right or wrong, we give here the modern description of a group. It is a nonempty set G on which any two elements a and b can be combined according to some specified operation $\circ$ having the following properties: (1) $a \circ (b \circ c) = (a \circ b) \circ c$ for all $a, b, c \in G$ (the associative law); (2) G contains an element e (the identity) such that $e \circ a = a \circ e = a$ for all $a \in G$; (3) for all $a \in G$ there exists an element a^{-1} (the inverse of a) such that $a \circ a^{-1} = a^{-1} \circ a = e$. These axioms can be weakened somewhat, but in practice there is no need to do so. If an object forms a group, it is usually possible to verify that fact quite easily using these axioms. Normally, the symbol $\circ$ for the group operation is omitted, and we write ab instead of $a \circ b$.

A good example of a group is the permutations of any set. This group is called the *symmetric group*. As we have already discussed this group, we give another example. The *general linear group* $GL(2, \mathbb{R})$ consists of 2×2 matrices of real numbers

$$a = \begin{pmatrix} a_{11} & a_{12} \\ a_{21} & a_{22} \end{pmatrix}$$

subject to the condition that the determinant $D(a) = a_{11}a_{22} - a_{12}a_{21}$ is not zero. The operation $\circ$ is matrix multiplication

$$ab = \begin{pmatrix} a_{11} & a_{12} \\ a_{21} & a_{22} \end{pmatrix} \begin{pmatrix} b_{11} & b_{12} \\ b_{21} & b_{22} \end{pmatrix} = \begin{pmatrix} a_{11}b_{11} + a_{12}b_{21} & a_{11}b_{12} + a_{12}b_{22} \\ a_{21}b_{11} + a_{22}b_{21} & a_{21}b_{12} + a_{22}b_{22} \end{pmatrix}.$$

In order for this set to be a group, the product, as we have defined it, must also belong to the group. Obviously the product ab is a 2×2 matrix. To show that its determinant is not zero, one can verify the identity $D(ab) = D(a)D(b)$.

The associative law can be verified, rather tediously. The identity for this group is

$$e = \begin{pmatrix} 1 & 0 \\ 0 & 1 \end{pmatrix}.$$

The inverse of the matrix a is

$$a^{-1} = \begin{pmatrix} \frac{a_{22}}{D(a)} & -\frac{a_{12}}{D(a)} \\ -\frac{a_{21}}{D(a)} & \frac{a_{11}}{D(a)} \end{pmatrix}.$$

One must be careful not to assume that $ab = ba$. This additional property (the commutative law) makes a group into a *commutative* group. If a group is being defined by specifying some rule $\circ$ and is required to be commutative as part of its definition, it is called an *abelian* group. This name seems to have arisen because Abel noticed an important case in which an equation can be solved algebraically. If an equation of degree n has one root x_n in terms of which the other roots can be expressed as rational functions $x_1 = f_1(x_n), \ldots, x_{n-1} = f_{n-1}(x_n)$, then the equation can be solved by radicals if $f_i(f_j(x)) = f_j(f_i(x))$ for all i and j.

The distinction between abelian and commutative groups is seldom important, and generally mathematicians use the term *abelian* to refer to any commutative group. When a group is commutative, its operation is usually written as addition rather than multiplication; that is, we write $a + b$ instead of ab. Neither of these operations should be regarded as the ordinary sum and product of numbers, although those are important examples of group operations. (The nonzero real numbers form a commutative group with multiplication as the operation. So do the positive real numbers, excluding zero.)

Incidentally, Klein and Lie thought that the inverse property of a group would follow from the cancellation law, as indeed it does when the permutations are on a finite set. But on an infinite set such is not the case. For example, on the set of positive integers, the mapping $a(n) = n^2$ obeys a cancellation law, since if $(b(n))^2 = (c(n))^2$ for all n and $b(n)$ and $c(n)$ are positive integers, then $b(n) = c(n)$ for all n. But the mapping a has no inverse. (It does have a "one-sided" inverse obtained by mapping n^2 to n and the nonsquare integers in any one-to-one manner whatsoever.)

As the work of Galois showed, a group may contain smaller groups using the same operation. These are called *subgroups*. The group S_3, which consists of all permutations of three letters $\{a, b, c\}$, for example, has six elements. It contains three two-element subgroups, each consisting of the identity element (which leaves every letter fixed) and one transposition, (ab) or (ac) or (bc). It also contains the subgroup whose elements are the identity and the two three-cycles (abc) and (acb). These are the even permutations, the so-called *alternating group* on three letters, usually denoted A_3.

As Galois noted, it is crucial whether a subgroup is *normal* or not. If a is an element of a normal subgroup and b is any element of the group (not necessarily in the subgroup), then $b^{-1}ab$ will also belong to the subgroup. Obviously, in a commutative group $b^{-1}ab = ab^{-1}b = ae = a$, so that all subgroups of a commutative group are normal. The alternating group A_3 is a normal subgroup of S_3. The other three subgroups, each having two elements, are not normal. In $GL(2, \mathbb{R})$, the subgroup of upper-triangular matrices of the form

$$\begin{pmatrix} a_{11} & a_{12} \\ 0 & a_{22} \end{pmatrix}$$

is not normal, as you can easily verify. The still smaller subgroup consisting of the multiples of the identity matrix, in which $a_{11} = a_{22} \neq 0$ and $a_{12} = 0 = a_{21}$, *is* normal.

Groups have permeated all of mathematics, and no mathematician can afford to neglect them. That is not surprising, since they are the essential tool for defining symmetry. Wherever there is symmetry in nature, whether in crystals or differential equations or anywhere else, *groups will arise!*

2. Rings

While abstract groups generalize the properties of permutations, abstract rings generalize the properties of the integers. The integers (positive, negative, and zero), together with the ordinary operation of addition, are a commutative group. But, as we know, the elements of this group can be multiplied also. The crucial property linking addition with multiplication is the *distributive property* expressed by the equation $a(b + c) = ab + ac$. A commutative group with the operation $+$ is called a *ring* if it has a second operation that is distributive with respect to addition.

The classical example of a ring, besides the integers and the Gaussian integers just mentioned, is the ring of polynomials in several variables, that is, polynomials like

$$p(x, y, z) = 5x^3 yz + 7yz^3 - 6xyz + 4x - 2y + 11 \,.$$

These actually form what is called an *algebra* since they are a vector space (see below). Multiplication of polynomials is both associative and commutative, that is, $p(qr) = (pq)r$ and $pq = qp$ for any polynomials p, q, and r.

The name *ring* was first introduced in German in 1914 by Adolf Fraenkel (1891–1965). The English word was introduced by Eric Temple Bell (1883–1960) in 1930. The name seems to have been inspired by the *cyclic* groups, whose elements can be multiplied by regarding them as the remainders when integers are divided by a given integer. (See Problems 1.1, 1.3, and 1.6 in Lesson 1.)

2.1. Associative rings. If the multiplication operation is also associative and/or commutative (as is the case with integers and polynomials), the ring is called an *associative* (and/or *commutative*) ring. Associative rings are one of the two most important classes of rings. A good example of an associative (but not commutative) ring is the complete set of 2×2 matrices with real entries. The addition is the obvious one:

$$a + b = \begin{pmatrix} a_{11} & a_{12} \\ a_{21} & a_{22} \end{pmatrix} + \begin{pmatrix} b_{11} & b_{12} \\ b_{21} & b_{22} \end{pmatrix} = \begin{pmatrix} a_{11} + b_{11} & a_{12} + b_{12} \\ a_{21} + b_{21} & a_{22} + b_{22} \end{pmatrix} \,.$$

The multiplication is the one defined above. We are no longer requiring the determinant to be nonzero; in fact, we *could not* impose that requirement, since the sum of two matrices, each with nonzero determinant, may have determinant zero.

Remark 11.4. The use of the same addition sign on both sides of this last equation is very convenient, indeed indispensable for sanity when working with matrices and vectors. However, *the two signs do not denote the same operation*. The sign on the left stands between two matrices and denotes matrix addition. The equation serves to tell us what matrix addition amounts to. Each addition sign on the right stands between two real numbers, which the reader, it is assumed, knows how to add. Please keep this conventional but ambiguous use of the notation for sums and products in mind throughout this section. It arises several times.

Ideals. For an associative ring, the analog of a normal subgroup is called an *ideal*. An ideal I is a subgroup of the additive group of the ring having the additional property that if $a \in I$ and b is any element of the ring, then ab and ba both belong to I. (That is a *two-sided* ideal. There are also left and right ideals, satisfying a weaker condition.) The whole point of ideals is that one can stratify a ring into disjoint cosets of an ideal, just as a group is stratified into disjoint cosets of a normal subgroup. These cosets can then be added and multiplied consistently by choosing a representative element from each coset that is to be added or multiplied, then adding or multiplying the representative elements, and finally taking the coset to which the sum or product belongs. The end result will be the same coset, no matter which representatives are chosen to determine it. For example, the even integers form an ideal in the ring of integers. In fact, the multiples of a fixed element always form an ideal, in any ring, called the *principal ideal* of that element.

2.2. Lie rings. The other major class of rings is formed by the *Lie rings*, which are usually vector spaces and are accordingly called *Lie algebras* (see below). In a Lie ring, the multiplication is usually written as a bracket $[a, b]$, and the associative law is replaced by the anticommutative property $[a, b] = -[b, a]$ and the *Jacobi identity*:

$$[a, [b, c]] + [b, [c, a]] + [c, [a, b]] = 0 .$$

A good example of a Lie algebra, for those who have studied calculus, is the algebra of vectors in three-dimensional space with the cross product as the multiplication. Given an associative ring, one can create a Lie ring on the same set by defining $[a, b] = ab - ba$. Notice that if the associative ring is commutative, the corresponding Lie ring is trivial (all products are zero).

Because of the wide variety of possible "multiplications," the theory of rings is much less unified and systematic than the theory of groups. (For example, one can form a trivial associative and commutative Lie ring out of any commutative group by simply defining $ab = 0$ for all a and b.) Two nonzero elements whose product is zero are called *zero divisors*. A commutative ring with identity but without any zero divisors is called an *integral domain.* The integers form an integral domain. The ring of 2×2 matrices does not, since multiplication is not commutative and there are zero divisors:

$$\begin{pmatrix} 1 & 0 \\ 0 & 0 \end{pmatrix} \begin{pmatrix} 0 & 0 \\ 0 & 1 \end{pmatrix} = \begin{pmatrix} 0 & 0 \\ 0 & 0 \end{pmatrix}.$$

2.3. Special classes of rings. A ring need not have any identity with respect to multiplication. (As one example, the even integers form a ring.) If a ring does have an identity for multiplication, the set of elements having a multiplicative inverse becomes important. These elements are called *units.* For the integers, the units are 1 and -1. For the ring of Gaussian integers, mentioned above, the units are 1, -1, i, and $-i$. The question of the factorization of the nonunits of a ring then becomes important.

If a nonunit in an integral domain cannot be written as a product of two or more nonunits, it is said to be *irreducible.* If a nonunit p has the property that it divides one of the two factors q and r whenever it divides their product qr, it is called a *prime.* In any ring, a prime is irreducible. If the converse is true, the ring is called a *unique factorization domain* or a *Gaussian domain.* As the name implies, in such a domain, there is essentially only one way to write an element as a product of irreducible elements. The integers are a Gaussian domain. That is the content of the "fundamental theorem of arithmetic," which asserts that every positive integer can be written in one and only one way as a product of positive primes. The Gaussian integers are another example. However, the set of complex numbers of the form $m + n\sqrt{-3}$ is *not* a Gaussian domain, since $2 \times 2 = (1 + \sqrt{-3}) \times (1 - \sqrt{-3})$.

3. Division rings and fields

If every nonzero element of a ring with identity has a multiplicative inverse, that ring is called a *division ring.* If, in addition, the multiplication is associative and commutative, the ring is called a *field.* We have already seen one example of a general division ring, the ring of quaternions, explored in Problems 1.7 through 1.10 and 6.4, and we have seen many examples of fields.

4. Vector spaces and related structures

Of all the abstract entities that mathematicians have invented and studied, vector spaces are among the simplest and also the most useful. To create a vector space, one needs first a commutative group, whose elements are to be called *vectors.* The simplest example is three-dimensional Euclidean space, denoted $\mathbb{R}^3$. It consists of triples of real numbers $\alpha = (a_1, a_2, a_3)$ with the group operation defined as the obvious addition: $\alpha + \beta = (a_1, a_2, a_3) + (b_1, b_2, b_3) = (a_1 + b_1, a_2 + b_2, a_3 + b_3)$.

Having the vectors, one also needs a field, whose elements are to be called *scalars.* In our example, that field will be the real numbers. Then, one must define the product of a vector and a scalar. In our example, the definition is obvious: $c\boldsymbol{\alpha} = c(a_1, a_2, a_3) = (ca_1, ca_2, ca_3)$. Again, notice that the multiplication on the left and in the middle is defined in terms of a different multiplication on the right, which the reader is assumed to know how to do.

The definition of scalar multiplication cannot be completely arbitrary. It must satisfy two distributive laws, one associative law, and a special "unitary" law. These are the following:

$$
\begin{aligned}
a(\boldsymbol{\alpha} + \boldsymbol{\beta}) &= a\boldsymbol{\alpha} + a\boldsymbol{\beta}, \\
(a + b)\boldsymbol{\alpha} &= a\boldsymbol{\alpha} + b\boldsymbol{\alpha}, \\
a(b\boldsymbol{\alpha}) &= (ab)\boldsymbol{\alpha}, \\
1\boldsymbol{\alpha} &= \boldsymbol{\alpha}.
\end{aligned}
$$

Although these seem obvious, they are independent requirements, and it is not difficult to construct examples in which exactly one of them fails. For example, by defining $c\boldsymbol{\alpha} = (ca_1, 0, 0)$, we would get the first three properties, but not the last.

An illustrative example of a vector space is provided by the space of polynomials in two variables with real coefficients. For example, if $p(x, y) = 3 + 5x - 2xy^2$ and $q(x, y) = -1 + 3x - y + xy - xy^2$, then the vector sum $p + q$ is defined as

$$(p + q)(x, y) = 2 + 8x - y + xy - 3xy^2,$$

and the scalar multiplication $5p$ is defined by $5p(x, y) = 15 + 25x - 10xy^2$.

Vector spaces have been mentioned in Lesson 4 and elsewhere, and their applications are almost infinite in number. Wherever records must be kept of different variables, vectors are apt to arise. In physics, this is obvious, where the three components of velocity, acceleration, force, torque, and the like are all conveniently represented as vectors. Economists likewise can use them to analyze the ways in which different sectors of an economic system interact. A good example of the latter is the *input–output analysis* of the Harvard economist Wassily Leontief (1906–1999).

The transformations of one vector space into another are extremely important, and these can be represented by matrices, making the whole subject of vector spaces, which mathematicians call *linear algebra*, very computational.

4.1. Modules. If the requirements for a vector space are relaxed slightly, so that the scalars are taken from a ring with a multiplicative identity instead of a field, the vector space is called a *module* (more strictly, a *unitary module*). Modules are relatively less important than vector spaces, but they have at least one important use in proving some crucial decomposition theorems for operators on a vector space. Mathematicians with an interest in string theory and superstring theory also make use of modules.

4.2. Algebras. If the elements of a vector space can be multiplied so as to satisfy the distributive and associative laws

$$\alpha(\beta + \gamma) = \alpha\beta + \alpha\gamma$$
$$c(\alpha\beta) = (c\alpha)(\beta),$$

the vector space is called an (associative) *algebra*. If the associative law is replaced by the anticommutative law and the Jacobi identity, it is called a *Lie algebra*. The polynomials in any number of variables with coefficients in a field form an associative algebra over that field. The vectors in $\mathbb{R}^3$, as already mentioned, form a Lie algebra with the cross product as the multiplication:

$$[\alpha, \beta] = \alpha \times \beta = (a_2b_3 - a_3b_2, a_3b_1 - a_1b_3, a_1b_2 - a_2b_1).$$

It may seem strange that the generic word for the entire subject that we are discussing is also applied to one very special object within that subject. The term is an American invention from the early twentieth century, and it was baffling to at least one or two European mathematicians. Salomon Bochner (1899–1982), who was in Oxford in 1925, heard a lecture by a young American woman on the concept of an algebra. He wrote in "Mathematical reflections" (*American Mathematical Monthly*, **83** (1974), No. 8) that, "She spoke in a well-articulated, self-confident manner, but none of us had remotely heard before the terms she used, and we were lost."

5. Conclusion

Although modern algebra has introduced hundreds of abstract structures, which are being studied by thousands of mathematicians, those mentioned above have been the most enduring. They are basic concepts that a technically trained person needs to know. The algebraic way of thinking gets into one's mind, in a manner that Klein did not expect when he disparaged abstract group theory. A great economy and unity is thereby produced. To take just one example, analysts had long known that functions of a real variable whose Fourier transforms never assume the value zero are of particular importance. Without going into details, we note that these functions enabled Norbert Wiener (1894–1964) to prove a very general theorem about convergence of series. When the subject of Banach algebras was created by Izrail' Moiseevich Gel'fand (b. 1913) in 1941, Wiener's theorem turned out to be a consequence of the fact that every ideal in the algebra of integrable functions is contained in a maximal ideal.

That example illustrates the basic principle of algebraic reasoning in analysis. Instead of studying an individual object or trying to construct an object having certain properties, one studies the whole class of objects having specified properties. If this class is an algebra, or a group, or a ring, it may be possible to show that the interrelationships among those objects, characterized by the algebraic laws the class obeys, imply that the desired object exists.

Having given this cursory glance at "what came after," we remind the reader that the story we mainly meant to tell was the story of the solution of polynomial equations. The present chapter exists because the end of that story in the midnineteenth century was not the end of algebra. Although the invention of equations was not a crime, and the study of polynomial algebra is not the punishment for it, an algebraist who lived through that crucial period might well be described by the final paragraph of Dostoevsky's *Crime and Punishment*:

> At this point a new story begins, the story of the gradual renewal of a man, the story of his gradual rebirth, his gradual passage from one world into another, his acquaintance with a new, heretofore completely unknown reality. That might form the subject for a new tale, but our present tale has reached its end.

Appendix: Some Facts about Polynomials

In the main portion of this book we have frequently alluded to, mentioned without proof, or otherwise abused the reader with, invocations of certain facts about polynomials that are well known to those who have had a course on modern algebra. To satisfy the curiosity or quell the frustration of readers who wonder what these things are all about, we list those facts here.

Rational roots of a polynomial with integer coefficients. *If a polynomial*

$$p(x) = m_0 x^n + m_1 x^{n-1} + \cdots + m_{n-1} x + m_n$$

with integer coefficients $m_0, \ldots, m_n$ has a rational root $r = s/t$, where s and t are integers with no common factors, then s divides m_n, and t divides m_0.

The proof of this fact is not difficult. We can write

$$0 = t^n p\left(\frac{s}{t}\right) = m_0 s^n + m_1 s^{n-1} t + \cdots + m_{n-1} s t^{n-1} + m_n t^n .$$

Since s divides every term here except the last one, it must also divide the last term. Just transpose $m_n t^n$ to the other side of the equation. Since s divides the right side, it must also divide the left side. However, since s has no factors in common with t^n, that means it must divide m_n. A similar argument shows that t must divide m_0.

A particular consequence of this last fact is that if $m_0 = 1$, as is frequently the case, then the rational roots of the equation are all integers, since $t = 1$. Thus, in particular, the equation $x^n - N = 0$ can have only integer solutions and irrational solutions. It follows that $\sqrt{2}$, $\sqrt{3}$, $\sqrt{5}$, and so on, are all irrational numbers.

This principle also converts the finding of rational roots of an equation with rational coefficients into a finite search. Hence that problem can be solved algorithmically, and it is always possible to tell when a polynomial with rational coefficients is irreducible over the rational numbers. We have invoked this principle, for example, in Lesson 11.

The Euclidean algorithm. Polynomials, like the integers, form a unique factorization domain. In fact, they are even better than that. They are a *Euclidean ring* since there is an algorithm for finding the greatest common divisor of any two elements. This algorithm consists of repeated divisions.

We will explain the procedure through an example, finding the greatest common divisor of 26,173,996,849 and 180,569,389.

We proceed as follows: $26, 173, 996, 849 \div 180, 569, 389$ is 144 with a remainder of 172,004,833. We then divide 180,569,389 by 172,004,833, getting a quotient of 1 and a remainder of 8,564,556. Next we divide 172,004,833 by 8,564,556, getting a quotient of 20 and a remainder of 713,713. We then divide 8,564,556 by 713,713 and get a quotient of 12 with no remainder, so that the greatest common divisor is 713,713. In other words, the last nonzero remainder is the greatest common divisor. This computation can be conveniently arranged with the divisions performed from right to left, so as to make the greatest common divisor appear at the extreme left.

$$
\begin{array}{c c c c}
12 & 20 & 1 & 144 \\
\overline{713713)8564556} & \overline{172004833)180569389} & & \overline{26173996849} \\
8564556 & 171291120 & 172004833 & 26001992016 \\
0 & 713713 & 8564556 & 172004833
\end{array}
$$

The proof that this procedure works is not difficult. The first division says that $26, 173, 996, 849 = 144 \times 180, 569, 389 + 172, 004, 833$. This equation shows that any number that divides both of the numbers 26,173,996,849 and 180,569,389 must also divide 172,004,833. (This is exactly the same reasoning used above when we were determining the possible rational roots of an equation with integer coefficients.) But then it follows from the equation for the next division, $180, 569, 389 = 172, 004, 833 + 8, 564, 556$, that it must also divide 8,564,556, and then that it must divide 713,713. At that point, we note that 713,713 does divide itself and 8,564,556, and therefore, since $172, 004, 833 = 8, 564, 556 \times 20 + 713, 713$, the number 713,713 also divides 172,004,833, and then it must divide 180,569,389, and finally 26,173,996,849. In other words, 713,713 *is* a common divisor of these two numbers, and *any* common divisor of them must also divide 713,713, which must therefore be the *greatest* common divisor.

If you have taken two years of algebra, you will have learned how to divide two polynomials and get a remainder that is "smaller" (of lower degree) than the divisor. Since the degree cannot decrease indefinitely, this same procedure, applied to two polynomials, will eventually lead to their greatest common divisor. For example, to find the greatest common divisor of $p(x) = x^5 - 3x^3 + 3x^2 - 10x + 6$ and $q(x) = x^4 + 3x^3 + 9x^2 + 6x + 14$, we first perform division with remainder:

$$p(x) = q(x)(x - 3) + r(x),$$

where $r(x) = -3x^3 + 24x^2 - 6x + 48$. We then write

$$q(x) = r(x)\left(-\frac{1}{3}x - \frac{11}{3}\right) + r_1(x),$$

where $r_1(x) = 95x^2 + 190$. At the next division, we find

$$r(x) = r_1(x)\left(-\frac{3x}{95} + \frac{24}{95}\right),$$

Since there is no remainder, $r_1(x) = 95(x^2 + 2)$, the last nonzero remainder, is the greatest common divisor. In this case, *greatest* means the polynomial of largest *degree* that divides both $p(x)$ and $q(x)$, since the division procedure that we use to find it works by decreasing the degrees of the remainders.

Since 95 is a unit in the ring of polynomials (its inverse is $\frac{1}{95}$), it doesn't really matter, and we could equally well say that $x^2 + 2$ is the greatest common divisor in this case.

This repeated division-with-remainder procedure, producing the greatest common divisor as the last nonzero remainder, is called the *Euclidean algorithm*. The name comes from Proposition 2 of Book 10 of Euclid's *Elements*: *If, when the smaller of two given quantities is continually subtracted from the larger, that which is left never divides evenly the one before it, the quantities are incommensurable.* In other words, if the algorithm fails to terminate in a finite number of steps, the two quantities in question have no common measure. In integral domains like the integers and the polynomials, where there is a positive-integer valued function of the quantities (the absolute value of the integers themselves in the case of the ring of integers, the degree of the polynomials in the case of the ring of polynomials) that takes a smaller value on the remainder than on the divisor, this algorithm *cannot* fail to terminate after a finite number of steps, since a strictly decreasing infinite sequence of positive integers is an impossibility. Integral domains with this property are called *Euclidean*.

Remark 1. The operations used in the Euclidean algorithm are all *rational* operations (no root extractions are involved). Hence the result is always a rational function of the data.

Remark 2. In the Euclidean algorithm, we use only the remainders and ignore the quotients. These quotients are not unimportant, however. They form the basis for the *continued-fraction* representation of the quotient of the two numbers under consideration.

Descartes' rule of signs. As shown above, for an equation with integer or rational coefficients, there is an algorithm for finding all rational roots. They must be of the form s/t, where s divides the constant term and t divides the leading coefficient.

For an equation with real coefficients, there is no such procedure for finding real roots. However, it is possible to set bounds on the number of real roots. The rule for doing so is as follows: *A polynomial $p(x)$ with real coefficients has no more positive roots than it has changes of sign.*

A *change of sign* is a pair of successive nonzero coefficients whose product is negative (in other words, one is positive, the other negative). Thus, we can conclude that the polynomial $x^7 + 3x - 1$ cannot have more than one positive root. In fact, it does have exactly one (between 0 and 1), since its value when $x = 0$ is negative and its value when $x = 1$ is positive. Similarly, $x^6 - x^3 + 1$ could not have more than two positive roots. In fact, it does not have any positive roots, or even any real roots, since it is $\left(x^3 - \frac{1}{2}\right)^2 + \frac{3}{4}$, which is always positive. Although the rule as we stated it sets only an upper bound on the number of roots, there are cases in which it is possible to get a lower bound as well. It is easy to see, for example, that a polynomial with exactly one change of sign must have exactly one positive root, since its sign at $x = 0$ is opposite to its sign for large positive values of x.

The proof of this rule can be found in older algebra books and is based on the simple idea that if $r > 0$, then $(x - r)p(x)$ has at least one more change of sign than $p(x)$ has. The proof is short, but almost impossible for the average student to understand. (Please take that last statement as a challenge!) In fact, Descartes stated his rule of signs in 1637, but the validity of his argument was disputed for about a century, until proofs similar to the one given below began to appear. A thorough study of the rule and its proofs was given by Gauss nearly two centuries later, in 1828. New proofs of this rule have continued to appear as recently as 2004 and 2006.

For students who have had a semester of calculus, the following proof will be more comprehensible than the algebraic proof given by Descartes. It is based on Rolle's theorem, named after Michel Rolle (1652–1719), which asserts that between any two zeros of a polynomial there must be a zero of its derivative. The derivative of a polynomial $p(x) = a_0 x^{n_0} + a_1 x^{n_1} + \cdots + a_{k-1} x^{n_{k-1}} + a_k$ (where $a_1, \ldots, a_k$ are nonzero real numbers and $n_0 > n_1 > \cdots > n_{k-1}$ are positive integers) is $p'(x) = n_0 a_0 x^{n_0-1} + n_1 a_1 x^{n_1-1} + \cdots + n_{k-1} a_{k-1} x^{n_{k-1}-1}$, . In other words, to produce $p'(x)$, you multiply each term in $p(x)$ by its exponent and decrease its exponent by 1. For the constant term a_k, which corresponds to x^0, you are multiplying by 0, and hence this term gets dropped. In proving this theorem, we may assume that the constant term a_k is not zero, since otherwise $p(x) = x^{n_{k-1}} q(x)$, where $q(x)$ is of lower degree and $p(x)$ and $q(x)$ obviously have exactly the same number of positive zeros and the same number of changes of sign.

Descartes' rule is obvious for linear polynomials. We can therefore proceed by induction on the degree of the polynomial $p(x)$ and use the fact that the degree of $p'(x)$ is one less than the degree of $p(x)$.

First note that $p'(x)$ has the same number of changes of sign as $p(x)$ if $a_{k-1} a_k > 0$ and one fewer if $a_{k-1} a_k < 0$. If $r_1 \leq r_2 \leq \cdots \leq r_m$ are the positive zeros of $p(x)$, Rolle's theorem guarantees that $p'(x)$ has positive zeros $s_1, \ldots, s_{m-1}$ satisfying

$$r_1 \leq s_1 \leq r_2 \leq s_2 \leq \cdots \leq s_{m-1} \leq r_m.$$

If $a_{k-1} a_k < 0$, this is all we need to finish the proof, since by induction $p'(x)$ must have at least $m - 1$ changes of sign and hence $p(x)$ must have at least m changes of sign. (Observe that this argument is valid even when multiple roots are counted according to their multiplicity.)

In the other case, since $a_k = p(0)$ and a_{k-1} is a positive multiple of the first nonzero derivative of $p(x)$ at $x = 0$, it follows (by calculus) that if $a_{k-1} a_k > 0$, then $p(x)$ must be either positive and increasing or negative and decreasing at zero, and therefore must achieve a maximum or minimum at some positive number s_0 less than its first positive zero r_1. It then follows that $p'(s_0) = 0$, and so $p'(x)$ has m positive zeros. Therefore $p'(x)$ has at least m changes of sign, and so $p(x)$ does also. The proof is now complete.

Negative roots. By the simple expedient of replacing x by $-x$, we can see that *a polynomial $p(x)$ with real coefficients has no more negative real roots than $p(-x)$ has changes of sign.*

Replacing x by $-x$ merely changes the sign of the terms corresponding to odd powers, of course. Thus, the polynomial $p(x) = x^2 - 3x + 1$ cannot have any negative roots, since $p(-x) = x^2 + 3x + 1$ has no changes of sign.

Answers to the Problems and Questions

Lesson 1

Problem 1.1. The solution of $x + 0 = 0$ is $x = 0$, which is $0 + 0$; the solution of $x + 1 = 0$ is $x = 1$, which is $1 + 0$; the solution of $x + 0 = 1$ is $x = 1$, which is $0 + 1$, and the solution of $x + 1 = 1$ is $x = 0$, which is $1 + 1$.

Problem 1.2. The solution of $x^2 + 1 = 0$ is $x = 1$, and this is a double root, since $(x + 1)^2 = x^2 + x + x + 1 = x^2 + 1$. The two solutions of $x^2 + x = 0$ are $x = 0$ and $x = 1$. Since $x^2 + x + 1 = 1$ for both $x = 0$ and $x = 1$, the equation $x^2 + x + 1 = 0$ has no solutions.

Problem 1.3. If $m = 3r + 1$ and $n = 3s + 2$, then $m + n = 3(r + s + 1)$, so that the remainder when $m + n$ is divided by 3 is 0. Similarly, if $m = 3r + 2$, then $m + n = 3(r + s + 1) + 1$, and so the remainder when $m + n$ is divided by 3 is 1.

Problem 1.4. Subtracting 1 means adding -1 and vice versa. Dividing by 1 (as usual) amounts to doing nothing. Dividing by -1 is the same as multiplying by -1. It leaves 0 fixed and interchanges 1 and -1.

Problem 1.5. The possible pairs of roots in the three-element field are $(-1, -1)$, $(0, 0)$, $(1, 1)$, $(-1, 0)$, $(-1, 1)$, and $(0, 1)$. The respective quadratic equations are

$$
\begin{aligned}
x^2 - x + 1 &= 0 \\
x^2 &= 0 \\
x^2 + x + 1 &= 0 \\
x^2 + x &= 0 \\
x^2 - 1 &= 0 \\
x^2 - x &= 0 .
\end{aligned}
$$

The only quadratic equations that have no solution in this field are $x^2 + 1 = 0$, $x^2 - x - 1 = 0$, and $x^2 + x - 1 = 0$.

Problem 1.6. $\frac{1}{2} = -2$ in this field, since $2 \times -2 = 1$.

Problem 1.7. 1. The identity

$$
\alpha \cdot \alpha = a_1^2 + a_2^2 + a_3^2
$$

is a straightforward computation.

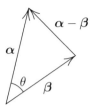

FIGURE 15. The vector law of cosines: $\boldsymbol{\alpha} \cdot \boldsymbol{\beta} = |\boldsymbol{\alpha}|\,|\boldsymbol{\beta}|\,\cos\theta$.

2. To prove
$$(\boldsymbol{\alpha} \cdot \boldsymbol{\beta})^2 \leq (\boldsymbol{\alpha} \cdot \boldsymbol{\alpha})\,(\boldsymbol{\beta} \cdot \boldsymbol{\beta}),$$
follow the hint, and divide the inequality $\boldsymbol{\gamma} \cdot \boldsymbol{\gamma} \geq 0$ by $\boldsymbol{\alpha} \cdot \boldsymbol{\alpha}$.

3. From the definition of the angle in terms of the dot product, perpendicularity means that the cosine of the angle between the two vectors is zero. Hence perpendicularity means that the dot product is zero. This "answer," however, is unsatisfying if the reader is not convinced that our definition of angle is intuitively correct. To see why it is, consider Fig. 15, which shows vectors $\boldsymbol{\alpha}$, $\boldsymbol{\beta}$, and $\boldsymbol{\alpha} - \boldsymbol{\beta}$. We take it as given that the correct geometric interpretation of the sum of two vectors is obtained from the head-to-tail juxtaposition familiar to physicists, so that $\boldsymbol{\alpha} - \boldsymbol{\beta}$ goes from the head of $\boldsymbol{\beta}$ to the head of $\boldsymbol{\alpha}$. (It is what you need to add to $\boldsymbol{\beta}$ to get $\boldsymbol{\alpha}$ as the sum.) We also assume that our interpretation of the length of a vector as the square root of the sum of the squares of its components is intuitively correct. With those assumptions, the law of cosines gives
$$|\boldsymbol{\alpha} - \boldsymbol{\beta}|^2 = |\boldsymbol{\alpha}|^2 + |\boldsymbol{\beta}|^2 - 2|\boldsymbol{\alpha}|\,|\boldsymbol{\beta}|\,\cos\theta.$$
Since the left side of this equation is
$$(\boldsymbol{\alpha} - \boldsymbol{\beta}) \cdot (\boldsymbol{\alpha} - \boldsymbol{\beta}) = |\boldsymbol{\alpha}|^2 + |\boldsymbol{\beta}|^2 - 2\boldsymbol{\alpha} \cdot \boldsymbol{\beta},$$
all we have to do is subtract the common terms from the two sides and divide by 2 to get the equation
$$\boldsymbol{\alpha} \cdot \boldsymbol{\beta} = |\boldsymbol{\alpha}|\,|\boldsymbol{\beta}|\,\cos\theta.$$

4. The remaining identities, namely
$$\begin{aligned}
\boldsymbol{\beta} \times \boldsymbol{\alpha} &= -\boldsymbol{\alpha} \times \boldsymbol{\beta}, \\
|\boldsymbol{\alpha} \times \boldsymbol{\beta}|^2 + (\boldsymbol{\alpha} \cdot \boldsymbol{\beta})^2 &= |\boldsymbol{\alpha}|^2|\boldsymbol{\beta}|^2, \\
\boldsymbol{\alpha} \times \boldsymbol{\beta} \cdot \boldsymbol{\alpha} &= 0, \\
\boldsymbol{\alpha} \times \boldsymbol{\beta} &= \pm|\boldsymbol{\alpha}|\,|\boldsymbol{\beta}|\sin\theta\boldsymbol{n}, \\
|\boldsymbol{\alpha} \times \boldsymbol{\beta}| &= |\boldsymbol{\alpha}|\,|\boldsymbol{\beta}|\sin\theta,
\end{aligned}$$
are all routine computations. (You may have to know a bit of linear algebra to deduce that the solution space of $\boldsymbol{\alpha} \cdot \boldsymbol{x} = 0 = \boldsymbol{\beta} \cdot \boldsymbol{x}$ is one-dimensional.)

Problem 1.8. As in the previous problem, the identity $(1+0)(a+\alpha) = a+\alpha$ is a routine computation.

Problem 1.9. Yet again, the identity $AB = BA$ if $A = a + \mathbf{0}$ is a routine computation.

Problem 1.10. With boring monotony, we find that the identity $A\bar{A} = a^2 + |\alpha|^2$ is another routine computation. The possible interpretations of $\frac{(0,0,3,0)}{(1,0,0,2)}$ are as $(0, 0, 3, 0)(1, 0, 0, 2)^{-1} = (0, -\frac{6}{5}, \frac{3}{5}, 0)$ or as $(1, 0, 0, 2)^{-1}(0, 0, 3, 0) = (0, \frac{6}{5}, \frac{3}{5}, 0)$.

Problem 1.11. The following ways of proceeding may occur to you. They are listed in increasing order of sophistication:

1. An experimental solution, for those who don't trust numbers and insist that only practical results are of value: Glue together some cheaper planks of the same dimensions and practice cutting the notches until you get it right.
2. A experimental paper-and-pencil method: Try different numbers until find two that add up to 19 and one is seven times the other.
3. Note that half of the plank must cover eight sides of the square, so divide 19 by 8. This will be the side of the square.
4. Solve the linear equation $x = 7(19 - x)$.
5. Solve the two linear equations $x + y = 19$, $y = 7x$.

Question 1.1. 1. To determine how much money you need to pay your bills, you add the bills. That is arithmetic.

2. If your current average over the first 60% of a course is 85%, the average x that you must maintain in order to get a semester average of 90% satisfies the equation $0.6 \times 0.85 + 0.4x = 0.9$. That is an algebra problem.

3. To use the formula $s = 4.9t^2$ to determine s given that $t = 7$, you substitute 7 for t in the equation. Since it is already solved for s, the computation you perform is arithmetic.

4. To determine the value of t given $s = 120$, you need to solve the equation for t, getting $t = \sqrt{s/4.9}$, and that is algebra.

Question 1.2. Suppose that there were a field consisting of six elements. One of the elements of this field is -1. Assume first that $-1 = 1$. Then $a = -a$ for all elements of the field. Let a be an element different from 0 and 1. Then the elements 0, 1, a, a^2, are all different. To see this, note that the equation $a^2 = 0$ implies $a = 0$. Likewise, the equation $a^2 = 1$ implies $a = 1$ or $a = -1 = 1$, and the equation $a^2 = a$ implies $a = 0$ or $a = 1$.

Now we cannot have $a^3 = 0$ or $a^3 = a$ or $a^3 = a^2$, since the first of these would imply $a = 0$ and the each of the other two would imply that either $a = 0$ or $a = 1$. Either a^3 is different from all four of 0, 1, a, and a^2, or $a^3 = 1$.

Let us assume temporarily that $a^3 = 1$. Then, since $a \neq 1$, we must have $a^2 + a + 1 = 0$, that is, $a^2 = a + 1$ and $a^2 + 1 = a$. In that case, the elements of the field must be $0, 1, a, a^2, b, c$, for some b and c different from 0, 1, a, and a^2 and from each other. Now b^2 must be different from 0, 1, a, a^2, and b. Indeed, $b^2 = 0$ implies $b = 0$, $b^2 = 1$ implies $b = 1$, $b^2 = a^2$

implies $b = a$, and $b^2 = b$ implies $b = 1$ or $b = 0$. Finally, $b^2 = a$ implies $(b^3)^2 = b^6 = (b^2)^3 = a^3 = 1$, and therefore $b^3 = 1$ also. That is, a and b are the solutions of $x^2 + x + 1 = 0$. But that means $ab = 1$, and therefore $b = a^2$, contrary to hypothesis. Therefore $c = b^2$, and the field consists of the distinct elements $0, 1, a, a^2, b, b^2$.

But now the product ab cannot be defined. It must be a nonzero element. If it is 1, then $ba = a^2 a$, and dividing by a yields $b = a^2$. If it is a, then $ba = a$, and again dividing by a yields $b = 1$. If it is a^2, then $ba = a^2$, and we get $b = a$. If it is b, dividing by b yields $a = 1$. Finally, if $ab = b^2$, then $a = b$.

Thus we conclude that 0, 1, a, a^2, and a^3 are all different. But then we cannot have $a^4 = 0$, $a^4 = 1$, $a^4 = a$, $a^4 = a^2$, or $a^4 = a^3$, since each of these implies either $a = 0$, or $a = 1$, or $a^2 = 1$, or $a^3 = 1$.

It follows that the elements of a six-element field in which $1 + 1 = 0$ would have to be $0, 1, a, a^2, a^3, a^4$, and $a^5 = 1$. Now we ask which element is $1 + a$. It is not 0, 1, a, or a^2, as already shown. Suppose that it is a^3. Then $1 + a^2$ must be a^4 (since there is nothing else left that it can be). But this implies that $a^4 = 1 + a^2 = (1 + a)^2 = (a^3)^2 = a^6 = a$, again a contradiction. Likewise, the assumption $1 + a = a^4$ leads to $1 + a^2 = a^3$, which implies $1 + a^3 = a^2$. But it also implies that $1 = a(1 + a^3)$, which in turn says that $1 + a^3 = a^4$, that is, $a^2 = a^4$, which is again a contradiction. Thus there is no six-element field in which $1 + 1 = 0$.

We now assume that $1 + 1 \neq 0$. For simplicity, we define $2 = 1 + 1$, $3 = 2 + 1 = 1 + 1 + 1$, $4 = 3 + 1 = 2 + 2 = 2 \cdot 2 = 1 + 1 + 1 + 1$, and so on. As we know, in the field of two elements $2 = 0$. We are now considering a hypothetical field of six elements in which $2 \neq 0$. We now temporarily assume $3 = 0$. That implies $4 = 1$, so that $2 \cdot 2 = 1$. Then the field contains the elements 0, 1, and 2, and a fourth element, which we call a. We cannot have $2a = 0$ or $2a = 1$ or $2a = 2$ or $2a = a$, since these imply respectively $a = 0$, $a = 2$ (because $2 \cdot 2 = 1$), $a = 1$, and $2 = 1$. Therefore the field consists of the five elements $0, 1, 2, a, 2a$ and a sixth element, which we label b. But then there is nowhere to go with $2b$. It cannot be a, since that implies $2a = 4b = b$ (because $3 = 0$), and all other hypotheses likewise lead to a contradiction. Thus we can safely exclude the possibility that $3 = 0$ also.

Now we certainly cannot have $4 = 0$, since that implies $2 \cdot 2 = 0$.

The next possibility is that $5 = 0$. Then $6 = 1$, i.e., $2 \cdot 3 = 1$, which in turn implies that $4 \cdot 4 = 1^2 = 1$ (since $9 = 4$). The field thus consists of $0, 1, 2, 3, 4, a$. Once again, there is simply nowhere to go with $2a$. The equations $2a = 0$, $2a = 1$, $2a = 2$, $2a = 4$, and $2a = a$ imply respectively $a = 0$, $a = 3$, $a = 1$, $a = 2$, and $2 = 1$. The remaining possibility $2a = 3$ implies that $4a = 1$ and hence $a = 4$. Thus we also rule out $5 = 0$.

Thus we must have $6 = 0$ in any such field but $2 \neq 0$ and $3 \neq 0$, and that is impossible, since $2 \cdot 3 = 6$.

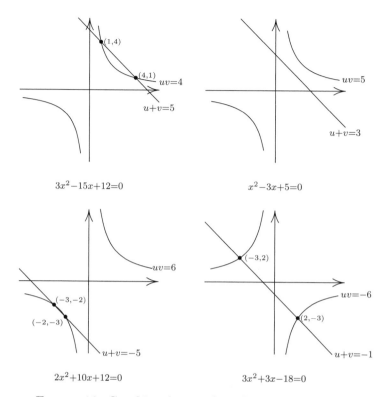

FIGURE 16. Graphic solution of quadratic equations.

Lesson 2

Problem 2.1. The graphs in Fig. 16 show that the roots of $3x^2 - 15x + 12 = 0$ are $x = 1$ and $x = 4$, that the equation $x^2 - 3x + 5 = 0$ has no real roots, that the roots of $2x^2 + 10x + 12 = 0$ are $x = -3$ and $x = -2$, and that the roots of $3x^2 + 3x - 18$ are $x = -3$ and $x = 2$.

Problem 2.2. We can solve by inspection. In the field of five elements, $x = 1$ and $x = 2$ are solutions of $x^2 + 2x + 2 = 0$. In the field of three elements, the polynomial $x^2 + 2x + 2$ assumes the values 2, 2, and 1, and hence has no zeros. (It is the same as $x^2 - x - 1$, which has no zeros, as shown in Problem 1.5.)

Problem 2.3. The system

$$
\begin{aligned}
x &+ 2y &- 3z &= 2, \\
2x &- 3y &+ 4z &= 1, \\
x &+ 9y &- 13z &= 5,
\end{aligned}
$$

is indeterminate. In fact, it has the general solution $x = (8 + z)/7$, $y = (3 + 10z)/7$, z arbitrary.

The system

$$
\begin{aligned}
x &+ y &+ z &= 5, \\
x &+ 2y &+ 3z &= 2, \\
x &+ 4y &+ 9z &= 3,
\end{aligned}
$$

is determinate. Its only solution is $x = \frac{23}{2}$, $y = -10$, $z = \frac{7}{2}$.

Problem 2.4. Combining the first equation with the second, third, and fourth respectively yields $y = (a - b)/2$, $z = (a - c)/2$, and $x = (a + d)/2$. The first equation then implies that $a - b - c + d = 0$. This is the consistency condition.

Problem 2.5. There is one solution: $m = 2$, $n = 3$. It was conjectured by Eugène Catalan (1814–1894) in 1844 that this is the only solution. That conjecture was finally proved by Preda Mihăilescu (b. 1955) in 2003.

Question 2.1. When $b < 0$, the hyperbola $uv = b$ occupies the second and fourth quadrants. Any straight line of negative slope ($u + v = -a$) must intersect each branch of the hyperbola in a point with one positive and one negative coordinate. This fact also follows since the polynomial $x^2 - ax + b$ has exactly one change of sign when $b < 0$. (See the discussion of Descartes' rule of signs in the Appendix.)

Question 2.2. For the equations $x^2 \pm 2ax + a^2 = 0$, the straight lines $u + v = \mp 2a$ are tangent to the hyperbola $uv = a^2$ at one vertex or the other.

Lesson 3

Problem 3.1. The quadratic formula that gives the solution of $ax^2 + bx + c = 0$ as $(-b \pm \sqrt{b^2 - 4ac})/(2a)$ is valid in any field where $2 \neq 0$. Of course, one needs to know which elements of the field have square roots, and adjoin a suitable square root if necessary, that is, if $b^2 - 4ac$ is not a square in the field. One can also go through the process of completing the square that leads to this formula, but again, it works only if $b^2 - 4ac$ is a square. Finally, one can simply substitute each element of the field in place of x and see whether the result is zero.

Problem 3.2. In the field of three elements we have $4 = 1$ and $2 = -1$, so that the solution of $ax^2 + bx + c = 0$ is

$$
x = \frac{b \pm \sqrt{b^2 - ac}}{a}.
$$

We must have $b^2 - ac = 0$ or $b^2 - ac = 1$ in order to take the square root, since -1 has no square root in this field. In that case, we can write $\sqrt{b^2 - ac} = b^2 - ac$, and the formula is even simpler.

In the field with five elements we have $4 = -1$, and so the formula is

$$x = \frac{-b \pm \sqrt{b^2 + ac}}{2a}.$$

Again, $b^2 + ac$ must be 0, 1, or -1, so that its square root can be taken.
  In both cases, the formula does work.

Question 3.1. In each problem, we are given information about the *result* of performing certain operations on a number (or numbers) and asked to find the number(s). In the Egyptian problem, the result of multiplying the unknown number by $\frac{8}{7}$ is 19. In the Mesopotamian problem, there are two unknowns (length and width), and we are told that their sum is 27, while their product plus their difference is 183. In the Chinese problem, the unknown is the number of hours required for the faster runner to overtake the slower, and we know that 40 times this number must equal 100. In al-Khwarizmi's inheritance problem, the unknown is a fictional amount of money to be repaid, and it must equal two-fifths of itself plus $3\frac{1}{2}$ *dirhems*. In the Japanese geometry problem, there are three unknowns. These are the three diameters of the different-sized circles, and we are given certain relations between them and the areas inside the largest and outside the three smallest, relations that are expressible as operations performed on the diameters.

Question 3.2. All the solutions given are formulaic (exact) except the geometry problem of Sawaguchi Kazuyuki.

Question 3.3. The quadratic formula does not work in the field of two elements, since $2 = 0$, and one cannot divide by 0. To solve the equation $x^2 + x + 1 = 0$ in this field, it is necessary to adjoin two more *cube* roots of unity, in addition to 1 itself.

Lesson 4

Problem 4.1. The identity

$$1 + 2 + \cdots + p = \frac{p(p+1)}{2}$$

is easily proved by induction. When $p = 1$, both sides equal 1. If the identity is true with $p = r$, then

$$1 + 2 + \cdots + r = \frac{r(r+1)}{2}.$$

Adding $r + 1$ to both sides then yields

$$1 + 2 + \cdots + r + (r+1) = \frac{r(r+1)}{2} + (r+1) = \frac{(r+1)(r+2)}{2},$$

which is the same identity with $p = r + 1$.
  Now working with the inequalities

$$\left(1 + 2 + \cdots + (n-1)\right)\frac{at^2}{n^2} < s < (1 + 2 + \cdots + n)\frac{at^2}{n^2}.$$

and invoking the given identity, we do indeed find that

$$\left(\frac{1}{2} - \frac{1}{2n}\right)at^2 < s < \left(\frac{1}{2} + \frac{1}{2n}\right)at^2.$$

Since these inequalities are true for all positive integers, we must have $s = \frac{1}{2}at^2$. For if $s < \frac{1}{2}at^2$, let $\varepsilon = \frac{1}{2}at^2 - s$, so that $\varepsilon > 0$. If we choose $n > at^2/(2\varepsilon)$, we then have $at^2/(2n) < \varepsilon$, and therefore $\frac{1}{2}at^2 - at^2/(2n) > \frac{1}{2}at^2 - \varepsilon = s$, which is a contradiction. A similar contradiction results from the assumption that $s > \frac{1}{2}at^2$.

Problem 4.2. The formula for s/t^2 is obtained by taking s as the distance from the point (r, vt) to the point $(r\cos(vt/r), r\sin(vt/r))$:

$$s^2 = r^2\left(1 - \cos\left(\frac{vt}{r}\right)\right)^2 + \left(vt - r\sin\left(\frac{vt}{r}\right)\right)^2.$$

A little trigonometry is then required to get rid of the cosine, namely the formula $1 - \cos\varphi = 2\sin^2\left(\frac{\varphi}{2}\right)$. You can then write this relation as

$$\frac{s^2}{t^4} = \frac{v^4}{4r^2}\left(\frac{\sin^4\left(\frac{vt}{2r}\right)}{\left(\frac{vt}{2r}\right)^4}\right) + \frac{v^4}{r^2}\left(\frac{r}{vt}\right)^2\left(1 - \frac{\sin\left(\frac{vt}{r}\right)}{\frac{vt}{r}}\right)^2.$$

Unfortunately, calculus is required to show that $(\sin\theta)/\theta \to 1$ and $(1/\theta)(1 - (\sin\theta)/\theta) \to 0$ as $\theta \to 0$. For that reason, we omit the proofs of these facts, even though the first of them is easy. Given those facts, you can see by letting t tend to zero that s^2/t^4 tends to $v^4/(4r^2)$, from which it follows that the instantaneous law of "falling" is $s = \frac{1}{2}(v^2/r)t^2$.

Comparing this law with the law for uniformly accelerated linear motion, we see that the magnitude of the acceleration must be $a = v^2/r$.

Problem 4.3. To express the linear velocity of the Moon in its revolutions around the Earth, we have

$$v = \frac{2\pi r}{T},$$

where $T = 27.3$ days $= 2.35872 \times 10^6$ seconds, and $r = 3.844 \times 10^8$ meters. Hence we find that $v = 1023.97$ meters per second. Actually, however, we don't need this number, since it is the acceleration v^2/r that we are interested in:

$$\frac{v^2}{r} = \frac{4\pi^2 r}{T^2} \approx 0.0027.$$

This is the value given in the text, and, as noted, it is approximately $1/3600$ of the acceleration of gravity at the surface of the earth. Hence if the acceleration of the Moon is indeed due to the Earth's gravity, that gravitational force must decrease in proportion to the square of the distance. (At least, that is the most elegant close fit for this single data point!)

Now, given that $v^2/r = C/r^2$, we find that $v = \sqrt{C}/\sqrt{r}$. We then note that the period T of a planet in a circular orbit satisfies

$$T = \frac{2\pi r}{v} = \frac{2\pi}{\sqrt{C}} r^{3/2} = K r^{3/2}.$$

Putting it another way, $T^2 = (4\pi^2/C)r^3$, which is Kepler's third law.

Problem 4.4. All these steps are reversible. If $T^2 = (4\pi^2/C)r^3$, then $v^2/r = C/r^2$. Hence, Kepler's law implies an inverse-square law of gravitation.

Problem 4.5. From the data given, we have

$$\frac{1}{\sqrt{\varepsilon_0 \mu_0}} = \frac{1}{\sqrt{3.536\pi}} \times 10^9 \text{ meters per second}.$$

Resorting to a calculator, we find that this is 3.00033×10^8 meters per second, almost precisely the speed of light. Given that this is the speed at which a self-sustaining electromagnetic wave *must* propagate, Maxwell drew the obvious conclusion that light is an electromagnetic wave.

Question 4.1. The similarity in mathematical form between the relations "area = length $\times$ width" and "distance = speed $\times$ time" (together with infinitesimal reasoning) leads to the conclusion that the area under the velocity curve is proportional to the distance traveled when the velocity is not uniform (acceleration is not zero). The particular case of constant acceleration was discussed early on and turned out to provide a good description of the motion of bodies falling near the Earth's surface.

Again, a mathematical form for the relation between distance and the square of the time leads to the expression v^2/r for the acceleration of a body in uniform circular motion. This expression linked the inverse-square law of gravitation to Kepler's third law, and supported the hypothesis that universal gravitation was responsible for both the orbits of the planets and the falling of bodies near the Earth's surface. The mere fact that the two laws (inverse-square relation for gravitation and Kepler's third law) were mathematically equivalent provided support for both. (Kepler's third law could be verified by observation, but the inverse-square law of gravitation could not.) The symbolic manipulation allowed by algebraic notation greatly facilitated the perception of this connection.

Measurements of charges moving under the influence of electricity and magnetism led to the determination of the electric permittivity and magnetic permeability of a vacuum. Maxwell's laws relating the electric and magnetic fields showed that an electromagnetic wave could sustain itself if it propagated at a speed determined by these two constants, and computation revealed that speed to be precisely the speed of light, thus unifying optical and electromagnetic phenomena.

None of this would have been possible without the formulas, expressed in algebraic notation and manipulated according to the rules of algebra and calculus. In order to reach our final results, we had to replace some expressions by others that were formally different but demonstrably equal to them. That is the essence of algebraic manipulation.

Question 4.2. This question is more philosophical than mathematical. There is something about the asymmetry of the equations that reconcile the observations of two observers in classical physics that is disquieting. If electric and magnetic fields are truly observable, why do different observers agree about the magnetic fields but not the electric ones? In relativity, they don't agree about either field, but at least the disagreement shows up symmetrically in the equations of transition between them. One is almost inclined to imitate Aristotle and assume that we have an intuitive feeling for the way a well-run universe would function. We could then say that *therefore* the relativistic equations must be the correct ones. Almost, but not quite. That kind of self-confidence has been shattered many times in the past, and the history of science is littered with the wreckage of elegant theories, such as the elastic-solid explanation of light propagation.

What role does mathematical elegance play in the acceptance of a physical theory? At the one extreme we have the scorn of Ludwig Boltzmann (1844–1906), who said, "Elegance is for tailors." At the other extreme, we have Henri Poincaré (1854–1912), who said, "If nature were not beautiful, it would not be worth knowing."

These two extremes reflect the attitudes of scientists, not the nature of the world itself. There is a tension between the "engineering" perspective that insists on dealing with "just the facts" and the "mathematical" perspective that insists (again to quote Poincaré) that "facts don't talk." Each has something to contribute. In these lessons, we are promoting the mathematical perspective sympathetically, showing the insight that can be gained by linking different, seemingly unrelated parts of the universe such as electricity, magnetism, and light, in a harmonious order. A very nice essay on the subject, by the physicist Norman David Mermin (b. 1935) can be read at the following website (as long as it stays available):

http://www.aip.org/pt/mar00/refmar.htm

Lesson 5

Problem 5.1. Let $p(x) = x^2 - 7$. Then $p(2) = -3$ and $p(3) = 2$, so the first digit is 2. Let $y = 2$. The array that gives the equation for y is

$$
\begin{array}{cccc}
1 & 1 & 1 & 1 \\
0 & 2 & 4 & 0 \\
-7 & -3 & 0 & 0
\end{array}.
$$

Thus we have $p_1(y) = y^2 + 4y - 3$. Since we are now into decimal places, we let $z = 10y$ and $q_1(z) = z^2 + 40z - 300$. Since $q_1(6) = -24$ and $q_1(7) = 29$, we now have the approximation $x = 2.6$ for the root. Let $z = w + 6$, and write

$$\begin{array}{cccc} 1 & 1 & 1 & 1 \\ 40 & 46 & 52 & 0. \\ -300 & -24 & 0 & 0 \end{array}$$

Thus, $p_2(w) = w^2 + 52w - 24 = 0$, and if $v = 10w$, we have $q_2(v) = v^2 + 520v - 2400$. Since $q_2(4) = -304$, while $q_2(5) = 225$, the root lies between 2.64 and 2.65.

Problem 5.2. Obviously the root is between 1 and 2. If we let $x = 1 + y$, the equation for y is found from the system

$$\begin{array}{ccccc} 1 & 1 & 1 & 1 & 1 \\ 0 & 1 & 2 & 3 & 0 \\ 0 & 1 & 3 & 0 & 0 \\ -3 & -2 & 0 & 0 & 0 \end{array}$$

so that $p_1(y) = y^3 + 3y^2 + 3y - 2 = 0$. Since we are now right of the decimal point, we let $z = 10y$, and write instead the equation $q_1(z) = z^3 + 30z^2 + 300z - 2000 = 0$, where we know that $0 < z < 10$. Now $q_1(4) = -256$, while $q_1(5) = 375$, so that the next digit of the solution is 4. That is, our next approximation to x is 1.4. Letting $z = 4 + w$, we get

$$\begin{array}{ccccc} 1 & 1 & 1 & 1 & 1 \\ 30 & 34 & 38 & 42 & 0 \\ 300 & 436 & 588 & 0 & 0. \\ -2000 & -256 & 0 & 0 & 0 \end{array}$$

Thus w satisfies $p_2(w) = w^3 + 42w^2 + 588w - 256 = 0$, and replacing w by $v = 10w$, we get $q_2(v) = v^3 + 420v^2 + 58800v - 256000 = 0$. Since $q_2(4) = -14016$ and $q_2(5) = 48625$, the next approximation to the root is 1.44. If you continue, you will get $x \approx 1.44225$.

Problem 5.3. The system that provides the equation for y is

$$\begin{array}{ccccc} 1 & 1 & 1 & 1 & 1 \\ 1 & 2 & -2 & -1 & 0 \\ 1 & -2 & 1 & 0 & 0. \\ 1 & -1 & 0 & 0 & 0 \end{array}$$

Here we use the fact that "3" is -2 and "4" is -1. Thus we have $y^3 - y^2 + y - 1 = 0$. Since the elements of this field are not ordered, there is no point in looking for a "change of sign" in the value of this polynomial. What this equation tells us is that $x = 1$ is not a solution, since the constant term here is not zero.

We can keep on guessing 1, however, since taking $y = 1$ amounts to taking $x = 2$. And obviously, this is a solution. The method will reveal that to us (as if it were not already obvious!). If we let $y = z + 1$, the equation for z will be derived from the system

$$
\begin{array}{rrrrr}
1 & 1 & 1 & 1 & 1 \\
-1 & 0 & 1 & 2 & 0 \\
1 & 1 & 2 & 0 & 0 \\
-1 & 0 & 0 & 0 & 0
\end{array}.
$$

Thus $y = 1$ $(x = 2)$ is a solution, and the equation satisfied by z is $z^3 + 2z^2 + 2z = 0$, that is, $z(z^2 + 2z + 2) = 0$. Hence either $z = 0$ or $z^2 + 2z + 2 = 0$. If you wish, you can solve this quadratic the same way, but obviously, its solutions are $z = 1$ and $z = 2$, meaning $x = -2$ or $x = -1$.

Question 5.1. We are assuming that $a_0 \neq 0$. Let $b_k = a_k/a_0$, $k = 1, 2, \ldots, n$. The roots of the given polynomial are the same as those of the polynomial $z^n + b_1 z^{n-1} + \cdots + b_{n-1}z + b_n$. We first invoke the "fundamental theorem of algebra" (discussed in Lesson 10) to argue that there is at least one root. Then we note that if $|z| \geq 1 + |b_1| + \cdots + |b_n|$, we have

$$
|b_1 z^{n-1} + \cdots + b_{n-1}z + b_n| < (|b_1| + \cdots + |b_n|)|z^{n-1}| < |z|^n ,
$$

and therefore

$$
|z^n + b_1 z^{n-1} + \cdots + b_{n-1}z + b_n| > |z^n| - |b_1 z^{n-1} + \cdots + b_{n-1}z + b_n| > 0 ,
$$

so that there are no zeros outside the given range. Hence all the zeros (and, in particular, at least one) must be in the indicated range.

Question 5.2. The most efficient way of finding the digit where the sign changes is repeated bisection. Try 5 first, then (depending on which sign is which) 3 or 8. In this way, you will not need to try more than four digits in order to get the answer.

Trying successively longer values of x instead of substituting y, z, and so on, in succession, leads to horrendously long and complicated decimal computations at a very early stage and is very much error-prone.

Question 5.3. The principal advantage of the Chinese method is the feature just mentioned: By working with an equation for a single-digit number at each stage, one avoids dealing with horrendously long decimal expansions. In terms of the actual number of steps required, the two methods are comparable.

Question 5.4. When the constant term in the equation for the new variable is zero, that new variable can be taken as 0, meaning that the previous equation was satisfied exactly by the value of the previous variable.

Lesson 6

Problem 6.1. The quadratic formula gives us

$$
x = \frac{i \pm \sqrt{-5 + 4i}}{2} .
$$

According to our formula, we have

$$\sqrt{-5+4i} = \sqrt{\frac{\sqrt{41}-5}{2}} + i\sqrt{\frac{\sqrt{41}+5}{2}}.$$

As a result, we have

$$x = \sqrt{\frac{-5+\sqrt{41}}{2}} + i\left(\frac{1}{2} + \sqrt{\frac{5+\sqrt{41}}{2}}\right)$$

or

$$x = -\sqrt{\frac{-5+\sqrt{41}}{2}} + i\left(\frac{1}{2} - \sqrt{\frac{5+\sqrt{41}}{2}}\right).$$

Problem 6.2. We are of course assuming $a \neq 0$. The quadratic formula reveals the roots to be

$$-\frac{b}{2a} \pm \sqrt{\left(\frac{b}{2a}\right)^2 - \frac{c}{a}}.$$

If one of these two roots has imaginary part zero, we must have

$$\mathrm{Im}\left(-\frac{b}{a}\right) = \pm\mathrm{Im}\sqrt{\left(\frac{b}{a}\right)^2 - \frac{4c}{a}}.$$

If we square both sides of this equation, we get one that is fully equivalent to it. Then, using our formula for the square root of a complex number, we find

$$-\frac{(\bar{a}b - a\bar{b})^2}{4|a|^4} = \frac{1}{2}\left(\left|\left(\frac{b}{a}\right)^2 - \frac{4c}{a}\right| - \mathrm{Re}\left(\left(\frac{b}{a}\right)^2 - \frac{4c}{a}\right)\right),$$

where $\bar{a}$ is, as usual, the complex conjugate of a. After expanding the real part on the right side here and transferring it over to the left side, we get an equation that can be written as

$$\mathrm{Re}\left(\left|\frac{b}{a}\right|^2 - \frac{4c}{a}\right) = \left|\left(\frac{b}{a}\right)^2 - \frac{4c}{a}\right|,$$

or, after multiplying through by $|a|^2$, as

$$\mathrm{Re}\left(|b|^2 - 4\bar{a}c\right) = \left|b^2 - 4ac\right|.$$

Problem 6.3. We have $u^2 = v$ and so $u^2 + u + 1 = v + u + 1 = 1 + 1 = 0$, and likewise $v^2 = u$, so that $v^2 + v + 1 = 0$. All four of the elements have square roots. (And u and v are cube roots of 1.)

Problem 6.4. If α is any vector of length r, that is, $|\alpha| = r$, then the quaternion $A = 0 + \alpha$ satisfies

$$A^2 = -\alpha \cdot \alpha = -r^2.$$

If $B = b + \beta$, then $B^2 = (b^2 - |\beta|^2) + 2b\beta$, and therefore we cannot have $B^2 + r^2 = 0$ if $b \neq 0$. Hence we have found all solutions of this equation. Notice that a quadratic equation in quaternions may have a simply enormous number of solutions—a whole two-dimensional sphere full of them—and all because quaternion multiplication is not commutative. If it were, the quaternions would be a field, in which a quadratic equation can have at most two solutions.

Question 6.1. If one of u and v is real and the other nonreal, then $u + v$ is nonreal. Hence b/a is nonreal. It follows that a and b cannot both be real.

Question 6.2. If the elements of a finite field are $a_1, \ldots, a_n$, then the polynomial

$$(x - a_1)(x - a_2) \cdots (x - a_n) + 1$$

has no roots in the field.

Lesson 7

Problem 7.1. If $y = q/(pw)$, then

$$
\begin{aligned}
0 &= y^3 + py + q \\
&= \frac{q^3}{p^3 w^3} + \frac{q}{w} + q \\
&= \frac{q}{w^3}\left(\frac{q^2}{p^3} + w^2 + w^3\right).
\end{aligned}
$$

Thus we have

$$w^3 + w^2 = N,$$

where $N = -q^2/p^3$.

Problem 7.2 Equations with only two terms: (1) $ax^3 = d$.

Equations with three terms: (2) $ax^3 = cx + d$, (3) $ax^3 + cx = d$, (4) $ax^3 + d = cx$, (5) $ax^3 = bx^2 + d$, (6) $ax^3 + bx^2 = d$, (7) $ax^3 + d = bx^2$.

Equations with all four terms: (8) $ax^3 = bx^2 + cx + d$, (9) $ax^3 + bx^2 = cx + d$, (10) $ax^3 + cx = bx^2 + d$, (11) $ax^3 + d = bx^2 + cx$, (12) $ax^3 + bx^2 + cx = d$, (13) $ax^3 + bx^2 + d = cx$, (14) $ax^3 + cx + d = bx^2$.

Problem 7.3. Obviously we need $a = B/A$, $b = \sqrt{C/A}$, $c = D/C$.

Now, given

$$
\begin{aligned}
xy &= bc \\
\left(x + \frac{a - c}{2}\right)^2 + (y - b)^2 &= \left(\frac{a + c}{2}\right)^2,
\end{aligned}
$$

we write $y = bc/x$, and substitute this value in the second equation. The result, after multiplying by x^2, is

$$x^2\left(x + \frac{a - c}{2}\right)^2 + b^2(c - x)^2 = x^2\left(\frac{a + c}{2}\right)^2.$$

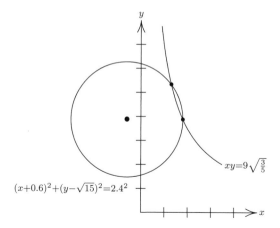

FIGURE 17. Graphical solution of $x^3 + 3x^2 + 15x = 27$.

Transposing the right side to the left and expanding, we get

$$x^4 + (a - c)x^3 - acx^2 + b^2(c - x)^2 = 0.$$

In other words

$$(x - c)(x + a)x^2 + (x - c)^2 b^2 = 0.$$

Hence, either $x = c$ or

$$x^3 + ax^2 + b^2 x = b^2 c.$$

Problem 7.4. Here we have $a = 3$, $b = \sqrt{15}$, $c = \frac{9}{5}$, so that the circle and hyperbola are given by

$$xy = 9\sqrt{\frac{3}{5}}$$

$$\left(x + \frac{3}{5}\right)^2 + (y - \sqrt{15})^2 = \left(\frac{12}{5}\right)^2,$$

Figure 17 shows the two curves, and seems to indicate that a solution can be found around $x = 1.3$. In fact, there is a solution close to $x = 1.30824$, as numerical methods will show.

Problem 7.5. For $x^3 + 153x - 4886 = 0$, we find

$$x = \sqrt[3]{2443 + \sqrt{51^3 + 2443^2}} - \sqrt[3]{2443 - \sqrt{51^3 + 2443^2}} = 17 - 3 = 14.$$

For the equation $x^3 - 6x^2 + 144x - 1539 = 0$, we first let $y = x - 2$, so that $x = y + 2$ and $y^3 + 132y - 1267 = 0$. We then find that

$$y = \sqrt[3]{\frac{1267}{2} + \sqrt{44^3 + \left(\frac{1267}{2}\right)^2}} - \sqrt[3]{\frac{1267}{2} - \sqrt{44^3 + \left(\frac{1267}{2}\right)^2}} = 11 - 4 = 7,$$

so that $x = 9$.

For $x^3 - x - 1 = 0$, we get

$$x = \sqrt[3]{\frac{1}{2} + \sqrt{-\left(\frac{1}{3}\right)^3 + \left(\frac{1}{2}\right)^2}} + \sqrt[3]{\frac{1}{2} - \sqrt{-\left(\frac{1}{3}\right)^3 + \left(\frac{1}{2}\right)^2}}$$

$$= \sqrt[3]{\frac{1}{2} + \sqrt{\frac{23}{108}}} + \sqrt[3]{\frac{1}{2} - \sqrt{\frac{23}{108}}} \approx 1.32472 \,.$$

For $x^3 - 6x^2 + 11x - 6 = 0$, the reduction $x = y + 2$ leads to the equation

$$y^3 - y = 0 \,.$$

The cubic discriminant is negative here. If we were mindless enough to apply the formula with $p = -1$, $q = 0$, we would get

$$y = \sqrt[3]{\sqrt{-\frac{1}{3}}} + \sqrt[3]{-\sqrt{-\frac{1}{3}}} \,.$$

Imitating Bombelli, we could take this to be zero, which is indeed a root of the equation. Of course, there is no need to do this, since it is obvious that $y = 0$, $y = -1$, or $y = 1$, and so $x = 1$, $x = 2$, or $x = 3$.

Problem 7.6. To say that α is a primitive pth root of unity, where p is a prime, is to say that $\alpha^k = 1$ if and only if k is a multiple of p. Then, for any j, $\alpha^{jk} = 1$ if and only if jk is a multiple of p. Since p is a prime, this is equivalent to saying that either j or k is a multiple of p; and since j is not a multiple of p, this means that jk is a multiple of p if and only if k is.

Problem 7.7. The fourth roots of unity are 1, -1, i, and $-i$. The primitive fourth roots of unity are i and $-i$.

Question 7.1. In the case of the square root, the relation between u, v, r, and s can be solved for u and v with an algebraic formula. This is done by eliminating v from the equation $2uv = s$, substituting its value in terms of u into the other equation, and then solving for u. The lucky thing is that v is a *rational* function of u and s in this one equation. In the case of the cubic equation, u and v are *algebraic* functions of each other and r and s, but not rational. Neither equation can be used to express v as a rational function of u, r, and s. In one equation we find $v = \sqrt{(u^3 - r)(3u)}$. The other is even worse, requiring us to solve a cubic equation for v.

Question 7.2. In order for the Cardano formula for the solution of $x^3 + px + q = 0$ to contain only rational functions of p and q, there must be *rational* numbers m and n such that $3mn = p$ and $m^3 - n^3 = q$. The solution of the equation is then $m - n$. In very many cases, an irrationality in m cancels a corresponding irrationality in n, leaving the solution $m - n$ rational, but expressed in terms of radicals.

Lesson 8

Problem 8.1. The Cardano formula gives

$$y = \sqrt[3]{-35 + \sqrt{35^2 - 13^3}} + \sqrt[3]{-35 - \sqrt{35^2 - 13^3}}$$

$$= \sqrt[3]{-35 + 18\sqrt{3}i} + \sqrt[3]{-35 - 18\sqrt{3}i} = 2\text{Re}\sqrt[3]{-35 + 18\sqrt{3}i}\,.$$

We need to make an inspired guess at a value of cube root. The best way to do that is to compute the root numerically using trigonometry. The number whose cube root we want can be written as

$$\sqrt{2197}(\cos\theta + i\sin\theta)\,,$$

where $\theta = \arccos\left(-35/\sqrt{2197}\right) \approx 2.4139$ (radians). Then the real part of one of the cube roots of this number is $(2197)^{1/6}\cos(0.80463) \approx 2.5$. Thus, the real part of this cube root is almost certainly $\frac{5}{2}$. That alone tells us that the root of the original equation should be $y = 5$, and we can verify that such is the case.

Problem 8.2. Although we don't need to know the imaginary part of this cube root in order to solve the original equation, we can now find it by solving the equation

$$\left(\frac{5}{2} + xi\right)^3 = -35 + 18\sqrt{3}i\,.$$

Since the real part of this equation is quadratic in x, obviously it is the part we should be looking at:

$$\frac{125}{8} - \frac{15}{2}x^2 = -35\,,$$

that is,

$$x^2 = \frac{27}{4}\,,$$

so that $x = \pm 3\sqrt{3}/2$. Computation verifies that indeed

$$\left(\frac{5}{2} + \frac{3\sqrt{3}}{2}i\right)^3 = -35 + 18\sqrt{3}i\,.$$

Problem 8.3. First we get rid of the cubic term by writing $y = x - \frac{3}{2}$, that is, $x = y + \frac{3}{2}$. The result is

$$y^4 = -\frac{1}{2}y^2 + 5y + \frac{91}{16} = 0\,.$$

We then add $2ty^2 + t^2$ to both sides, getting

$$(y^2 + t)^2 = \left(2t - \frac{1}{2}\right)y^2 + 5y + \left(t^2 + \frac{91}{16}\right)\,.$$

The condition for the right side to be a perfect square is

$$25 - 4\left(2t - \frac{1}{2}\right)\left(t^2 + \frac{91}{16}\right) = 0\,,$$

that is,

$$8t^3 - 2t^2 + \frac{91}{2}t - \frac{291}{8} = 0.$$

Fortunately, $t = \frac{3}{4}$ is one of the solutions of this resolvent cubic equation. Our original quartic therefore breaks up into two equations

$$y^2 + \frac{3}{4} = \pm\left(y + \frac{5}{2}\right),$$

which yield $y = \frac{1}{2} \pm \sqrt{2}$ and $y = -\frac{1}{2} \pm \sqrt{3}i$, as solutions. Hence $x = y + \frac{3}{2} = 2 \pm \sqrt{2}$ and $x = 1 \pm \sqrt{3}i$.

Problem 8.4. Viète's method converts the equation $y^3 + py + q = 0$ into $z^3 - \frac{3}{4}z = (3\sqrt{3}q)/(8p\sqrt{-p})$ by means of the substitution $y = \sqrt{-4p/3}z$. In the present case, that gives us the equation

$$z^3 - \frac{3}{4}z = -\frac{3\sqrt{3}(5\sqrt{6} + 6\sqrt{5})}{8(11 + \sqrt{30})\sqrt{11 + \sqrt{30}}}.$$

The right side must represent $\frac{1}{4}\cos\theta$. Hence we can express our answer as $z = \cos(\theta/3)$, where

$$\theta = \arccos\left(-\frac{3\sqrt{3}(5\sqrt{6} + 6\sqrt{5})}{2(11 + \sqrt{30})\sqrt{11 + \sqrt{30}}}\right).$$

Any calculator will tell you that $\theta = 3.06272$ radians, and adding 2π and 4π gets two other values of θ, namely, 9.3459 radians and 15.6291 radians. After dividing by 3 and taking the cosine, we get the values of z: 0.522594, -0.999654, and 0.477061. These correspond to y values of 2.44949, -4.68556, 2.23607. Those with a sensitivity to common square roots will recognize that the first of these represents $\sqrt{6}$ to the given accuracy and the last represents $\sqrt{5}$. Since the squared term is missing, the third root must be $-(\sqrt{6} + \sqrt{5})$. It is easy to verify that these are indeed the roots.

Problem 8.5. We begin with the equations

$$0 = y^3 + 18y + 30,$$
$$z = y^2 + ry + s.$$

Multiplying the second equation by y and subtracting the first, we obtain the equation

$$zy = ry^2 + (s - 18)y - 30,$$

which we can write as

$$y(z + 18 - s) = ry^2 - 30 = r(z - ry - s) - 30.$$

and conclude that

$$y = \frac{rz - 30 - rs}{z + 18 + r^2 - s}.$$

When this value is substituted into the equation $y^3 + 18y + 30 = 0$ and the denominator is cleared, the result is the equation

$$(r^3 + 18r + 30)(z^3 + (36 - 3s)z^2 + (18r^2 + 90r + 3s^2 - 72s + 324)z$$
$$+ 30r^3 - 18r^2s - 90rs + 540r - s^3 + 36s^2 - 324s - 900) = 0.$$

Thus, either r is a solution of the original equation, or

$$z^3 + (36 - 3s)z^2 + (18r^2 + 90r + 3s^2 - 72s + 324)z$$
$$+ 30r^3 - 18r^2s - 90rs + 540r - s^3 + 36s^2 - 324s - 900 = 0.$$

Choosing $s = 12$ will cause the coefficient of z^2 to vanish. Then, solving the equation

$$18r^2 + 90r - 108 = 0,$$

that is,

$$r^2 + 5r - 6 = 0,$$

for r and inserting this value of r will cause the coefficient of z to vanish, leaving a "pure" equation. This gives us two choices for r: $r = -6$ and $r = 1$. Obviously, the latter is simpler, and leads to

$$z^3 = 2058.$$

We then have the equation

$$y^2 + y + (12 - 7\sqrt[3]{6}) = 0.$$

and so

$$y = \frac{-1 \pm \sqrt{28\sqrt[3]{6} - 47}}{2}.$$

Actually, only the negative sign in front of the square root yields a solution of the original equation. If you have the patience, you can verify that with this value of y, which is approximately -1.484806656062487, you get

$$y^3 + 18y + 30 = -\frac{7}{2}\left((11 - 3\sqrt[3]{6}) + (1 + \sqrt[3]{6})\sqrt{28\sqrt[3]{6} - 47}\right).$$

It is then straightforward to verify that

$$(11 - 3\sqrt[3]{6})^2 = (1 + \sqrt[3]{6})^2(28\sqrt[3]{6} - 47).$$

Hence this value really is a solution.

If we had used the Cardano formula to solve this equation, which, by Descartes' rule of signs (see the Appendix), has only this one (negative) real root, we would have obtained an alternative expression for it:

$$y = \sqrt[3]{6} - \sqrt[3]{36}.$$

You can verify easily that

$$\left(\frac{1}{2} + \sqrt[3]{6} - \sqrt[3]{36}\right)^2 = \frac{28\sqrt[3]{6} - 47}{4}.$$

As just noted, the value $r = 1$ leads to the two values $y = \sqrt[3]{6} - \sqrt[3]{36} \approx -1.48481$ and $y = -1 - \sqrt[3]{6} + \sqrt[3]{36} \approx 0.48481$. The second is impossible, since this equation has no real positive solutions.

We could have used $r = -6$ also, leading to $z = 7\sqrt[3]{36}$. In that case, we would get the two values $y = \sqrt[3]{6} - \sqrt[3]{36} = -1.48481$ and $y = 6 - \sqrt[3]{6} + \sqrt[3]{36} \approx 7.48481$. Again, the second value is impossible, since the equation has no real positive solutions.

Even without invoking Descartes' rule of signs, we could have simply evaluated $y^3 + 18y + 30$ for numerical values of the roots we found sufficiently precise to show us which one is correct, and which is extraneous.

Problem 8.6. This time, we'll skip all the explanation and go directly to the formulas. We know that $s = 2p/3 = 24$ in this case, and we find r from the equation

$$36r^2 - 36r - 432 = 0 \,,$$

which is to say that

$$r^2 - r - 12 = 0 \,.$$

Hence $r = 4$ or $r = -3$. Let's keep things simple and just use $r = 4$. We then get z from the equation $z^3 = 16464$; that is, $z = 14\sqrt[3]{6}$. We then find y by solving

$$y^2 + 4y + (24 - 14\sqrt[3]{6}) = 0 \,.$$

The result is

$$y = 2\sqrt[3]{6} - \sqrt[3]{36} \quad \text{or} \quad y = -4 - 2\sqrt[3]{6} + \sqrt[3]{36} \,.$$

All we need is sign rules to see that the first of these must be correct. The quadratic equation has one positive root and one negative root, whereas the original cubic has only a positive root. Hence we must choose the first of these possible values, $y = 2\sqrt[3]{6} - \sqrt[3]{36}$.

Problem 8.7. We have the pair of equations

$$\begin{aligned} 0 &= y^5 + by^3 - cy^2 + dy - e \,, \\ z &= y^4 - py^3 + qy^2 - ry + s \,. \end{aligned}$$

Multiplying the second by y and subtracting the first yields

$$\begin{aligned} zy &= -py^4 + (q - b)y^3 - (r - c)y^2 + (s - d)y + e \\ zy &= (-p^2 + q - b)y^3 + (pq - r + c)y^2 - (pr - s + d)y - p(z - s) + e \,. \end{aligned}$$

Thus we do get the equation

$$0 = y^3 + \frac{pq - r + c}{q - p^2 - b}y^2 + \frac{s - pr - d - z}{q - p^2 - b}y + \frac{e + ps - pz}{q - p^2 - b} \,,$$

and this is indeed a cubic equation in y, enabling us (in principle) to express y in terms of z. However, in contrast to the case of the Tschirnhaus solution of the cubic, y is not a *rational* function of z. Substituting the expression for y in terms of z into the polynomial does not lead to a polynomial equation in z, but rather one that involves z under a radical. If the radical is removed by symmetrizing, the resulting polynomial equation will be of degree larger than 5.

Problem 8.8. The rational substitution $y = x + \frac{1}{4}$ leads to an equation

$$y^4 = ay^2 + by + c,$$

where a, b, and c are rational numbers. Forming the resolvent cubic, we get the two equations

$$y^2 + t = \pm\left(\sqrt{a + 2t}\,y + \frac{b}{2\sqrt{a + 2t}}\right),$$

where t satisfies $4(2t + a)(t^2 + c) - b^2 = 0$, and hence can be expressed in terms of a, b, and c, using only square and cube roots. The solutions y can then be expressed in terms of square roots of these quantities.

Problem 8.9. The trigonometric identity is a result of the fundamental addition formula $\cos(a + b) = \cos a \cos b - \sin a \sin b$, which in the case when $a = b$ yields $\cos 2a = \cos^2 a - \sin^2 a = 2\cos^2 a - 1$. Similarly, since $\sin(a+b) = \sin a \cos b + \cos a \sin b$, we get $\sin 2a = 2 \sin a \cos b$, and from that

$$\cos 3a = \cos(2a + a) = \cos 2a \cos a - \sin 2a \sin a =$$
$$= 2\cos^3 a - \cos a - 2\sin a \cos a \sin a = 4\cos^3 a - 3\cos a.$$

Next,

$$\sin 3a = \sin 2a \cos a + \cos 2a \sin a$$
$$= 2 \sin a \cos^2 a + (2\cos^2 a - 1)\sin a = (4\cos^2 a - 1)\sin a.$$

We then have

$$\cos 5\theta = \cos(3\theta + 2\theta)$$
$$= \cos 3\theta \cos 2\theta - \sin 3\theta \sin 2\theta$$
$$= (4\cos^3\theta - 3\cos\theta)(2\cos^2\theta - 1) - (4\cos^2\theta - 1)(2\cos\theta)\sin^2\theta$$
$$= 8\cos^5\theta - 10\cos^3\theta + 3\cos\theta - 10\cos^3\theta + 2\cos\theta$$
$$= 16\cos^5\theta - 20\cos^3\theta - 10\cos^3\theta + 5\cos\theta$$
$$= (\cos\theta - 1)(4\cos^2\theta + 2\cos\theta - 1)^2 + 1.$$

Taking $\theta = 2\pi/5$ here, we have $\cos 5\theta = 1$, and therefore, since $\cos\theta - 1 \neq 0$,

$$4\cos^2\theta + 2\cos\theta - 1 = 0.$$

This equation has only one positive solution:

$$\cos\theta = \frac{-2 + \sqrt{20}}{8} = \frac{-1 + \sqrt{5}}{4}.$$

Since $2\pi/5 < \pi/2$, we have

$$\sin\theta = \sqrt{1 - \cos^2\theta} = \sqrt{\frac{5 + \sqrt{5}}{8}}.$$

It thus follows that one of the fifth roots of unity is

$$\frac{-1+\sqrt{5}}{4}+\sqrt{\frac{5+\sqrt{5}}{8}}i\,.$$

Problem 8.10. We have the two equations

$$x^2-4=\pm\sqrt{2}i(x-2)\,.$$

Thus we need to solve

$$x^2-\sqrt{2}ix+2\sqrt{2}i-4=0$$

and

$$x^2+\sqrt{2}ix-(2\sqrt{2}i+4)=0\,.$$

That gives us the four solutions

$$x=\frac{\sqrt{2}i\pm\sqrt{14-8\sqrt{2}i}}{2}$$

and

$$x=\frac{-\sqrt{2}i\pm\sqrt{14+8\sqrt{2}i}}{2}\,.$$

Since $\sqrt{14+8\sqrt{2}i}=\pm(4+\sqrt{2}i)$, the root is either 2 or $-2-\sqrt{2}i$.

Problem 8.11. The elimination procedures are trivial. Notice that the equation $r(27q^2+4p^3)=0$ would be satisfied if the discriminant were zero (which means that the original equation has a multiple root), or if $r=0$. However, if $r=0$, then $p=0$ also, because of the equation that determines r. That in turn transforms the original equation into a pure equation $y^3+q=0$, whose solution is again trivial.

Question 8.1. A double root of the polynomial $p(x)=a(x-r)^2(x-s)=ax^3-a(s+2r)x^2+(ar^2+2ars)x-ar^2s$ is also a root of its derivative $p'(x)=a(2(x-r)(x-s)+(x-r)^2)=a(x-r)(3x-2s-r)$, which is a *quadratic* polynomial.

Lesson 9

Problem 9.1. The value of $tu+vw$ is completely determined by choosing one of u,v,w to be the coefficient of t. Hence it has only three values. Consider the equation

$$(z-tu-vw)(z-tv-uw)(z-tw-uv)=0\,.$$

By dint of tedious computation (or by invoking a computer algebra program), we find that this equation can be rewritten as

$$z^3-bz^2+(ac-4d)z+(4bd-c^2-a^2d)=0\,.$$

Now consider the system

$$t + u + v + w = a$$
$$tu + vw = g_1$$
$$tv + uw = g_2$$
$$tw + uv = g_3.$$

We have $(t+w)+(u+v) = a$ and $(t+w)(u+v) = (tu+vw)+(tv+uw) = g_1 + g_2$. Thus, for the two quantities $(t+w)$ and $(u+v)$, we know their sum and product. That, as we have emphasized repeatedly, amounts to knowing the coefficients of the quadratic equation having these two quantities as roots. Thus $t+w$ and $u+v$ are the roots of the equation $y^2 - ay + (g_1 + g_2) = 0$, and so they are

$$\frac{a \pm \sqrt{a^2 - 4(g_1 + g_2)}}{2}.$$

Thus, with suitable choices of the square roots in each case, we must have

$$t + w = \frac{a + \sqrt{a^2 - 4(g_1 + g_2)}}{2},$$

$$u + v = \frac{a - \sqrt{a^2 - 4(g_1 + g_2)}}{2},$$

$$t + u = \frac{a + \sqrt{a^2 - 4(g_2 + g_3)}}{2},$$

$$v + w = \frac{a - \sqrt{a^2 - 4(g_2 + g_3)}}{2}.$$

Problem 9.2. The sum of the first two equations minus the sum of the last two equations is identically zero (on both sides). In other words, the fourth equation can be derived from the first three by adding the first two and subtracting the third.

Problem 9.3. If $tu + vw = g_1$ and $d = tuvw$, then tu and vw are the two roots of the quadratic equation

$$x^2 - g_1 x + d = 0.$$

We now know tu and $t+u$, and hence can solve one more quadratic equation to find t and u.

Problem 9.4. We can categorize the permutations of five letters by the number of points left fixed:

Five fixed points. This must be the identity permutation, which leaves every element where it was. There is one such permutation.

Four fixed points. There are no permutations that leave exactly four elements fixed, since any such permutation must also leave the fifth element fixed.

Three fixed points. Such a permutation is a simple 2-cycle (transposition). There are 10 of these, since there are 10 ways of choosing the three fixed points (or the two nonfixed points).

Two fixed points. Once the two fixed elements are chosen (which can be done in 10 ways, as we know), there are two 3-cycles that move all three of the other elements. Hence there are 20 of these.

One fixed point. The one fixed point can be chosen in five ways. After that, there are six 4-cycles that move all of the remaining elements, and three pairs of 2-cycles that also move all of the remaining elements. Hence, there are 45 of these.

No fixed points. These must be either 5-cycles (and there are 24 of those) or a 3-cycle and a 2-cycle. For each fixed choice of the elements of the 3-cycle, there are two cyclic permutations that move all three elements. Hence there are 20 of the latter, for a total of 44 permutations with no fixed points.

Thus, we find the following categories of permutations: (1) the identity, (2) a single 2-cycle, (3) a single 3-cycle, (4) a single 4-cycle, (5) a single 5-cycle, (6) a pair of disjoint 2-cycles, (7) a disjoint 2-cycle and 3-cycle.

Problem 9.5. Obviously, the order of a cycle is its length (the number of elements it contains). Since disjoint cycles commute, we have $(\sigma\tau)^n = \sigma^n\tau^n$, and the only way this last permutation can leave every element fixed is for both σ^n and τ^n to do so. Hence n must be a multiple of the order of both σ and τ.

Problem 9.6. A permutation of four symbols must be one of the following: (1) the identity (order 1), (2) a single transposition (order 2), (3) a single 3-cycle (order 3), (4) a 4-cycle (order 4), and (5) a pair of disjoint 2-cycles (order 2). Hence the possible orders are 1, 2, 3, and 4.

Problem 9.7. From the list of permutation types compiled above we obtain the following: (1) the identity (order 1), (2) a single 2-cycle (order 2), (3) a single 3-cycle (order 3), (4) a single 4-cycle (order 4), (5) a single 5-cycle (order 5), (6) a pair of disjoint 2-cycles (order 2), and (7) a disjoint 2-cycle and 3-cycle (order 6). Thus, only the 5-cycles have order 5.

Problem 9.8. We have already shown that a 5-cycle σ has order 5, and in any case, this is quite obvious. Since f assumes fewer than five values, we must have $f \circ \sigma^i = f \circ \sigma^j$ for some integers satisfying $1 \le i < j \le 5$, and hence $f \circ \sigma^{j-i} = f$. So f is invariant under σ^{j-i}. But then it must also be invariant under $\sigma^{k(j-i)}$, $k = 1, 2, 3, 4, 5$, and these permutations are simply $\sigma, \sigma^2, \ldots, \sigma^5$ (the identity), in some order. In particular, since σ must be among them, f must be invariant under σ, that is, under any 5-cycle.

Problem 9.9. Let the original ordering be (s, t, u, v, w). The effect of $(suwtv)$ is to produce the ordering (v, w, s, t, u). Then the effect of $(wuvts)$ is to produce (u, s, t, v, w), that is, the two together have the same effect that (stu) has.

Problem 9.10. It follows immediately that $f_1 \circ (st) = f \circ (st) \circ (st) = f$, and obviously f_1 is also invariant under all 5-cycles, and hence also under all 3-cycles. Then $f \circ (tu) = f_1 \circ (st) \circ (tu) = f_1 \circ (stu) = f_1$. Thus, if $f_1 = f$, then f assumes only one value (is symmetric); otherwise f assumes only two values. This is the case in particular if it assumes fewer than five values.

Problem 9.11. Write $f \circ (stuv) = f \circ (stu)(uv) = f \circ (uv)$. Then $f \circ (stuv)^3 = f \circ (svut) = f \circ (tsvu) = f \circ (vu) = f \circ (uv) = f \circ (stuv)$.

Problem 9.12. Suppose that a configuration is $a, \ldots, p, q, r, s,$ "16", $t, u, v, w, \ldots,$ z (reading from left to right and top to bottom). The allowable moves of "16" are an interchange with p (moving "16" up a row), s (moving "16" left by one column), t, or w. Each of these moves switches the color of the square on which "16" is located. If the interchange is with p, the number of inversions involving "16" increases by four, while the number involving p and one of q, r, and s switches from (say) n to $3 - n$, in other words, the change is $3 - 2n$, which is an odd number, so that the total change is $7 - 2n$, also an odd number. If the interchange is with s, the total number of inversions increases by one. If it is with t, the total number decreases by one. Finally, if it is with w, the change is $-1 - 2n$, again an odd number. Thus, every allowable move changes the color of the blank square and changes the total number of inversions by an odd number. The total number is therefore odd if the blank square is white and even if it is shaded.

Problem 9.13. In this case, we can ignore the inversions involving the fictional "25." Every move changes the number of inversions involving the numbers $1, \ldots,$ 24 by an even number. (There is no change when a square is moved left or right. When one is moved up or down, the total change is from x to $4 - x$, a total change of $4 - 2x$, or from x to $6 - 2x$, both of which are even numbers. Thus, no move of any kind will change the parity of the number of inversions in the numbered tiles.)

Question 9.1. Since every permutation is a composition of transpositions, there can be only two values altogether.

Question 9.2. The function $f(s, t, u, v, w) = s + 2t + 3u + 4v + 5w$ assumes 120 formally different values.

Question 9.3. The usual elimination procedure for solving a pair of linear equations leads to

$$a = \frac{v - \alpha u}{\alpha^2 - \alpha} = \frac{1}{3}(\alpha - \alpha^2)(v - \alpha u) = \frac{u + \alpha v + \alpha^2(-u - v)}{3}.$$

Lesson 10

Problem 10.1 Suppose $p(x, y) = p_0(x) + p_1(x)y + \cdots + p_{n-1}(x)y^{n-1} + p_n(x)y^n$, where $p_n(x)$ is not identically zero. As shown in Question 5.1

above, if $p(x, f(x)) \equiv 0$, then

$$|f(x)| < 1 + \left|\frac{p_0(x)}{p_n(x)}\right| + \cdots + \left|\frac{p_{n-1}(x)}{p_n(x)}\right|.$$

Let k be the maximum of $\deg p_j(x) - \deg p_n(x)$. Then the preceding inequality implies that

$$\frac{f(x)}{x^{k+1}} \to 0$$

as $x \to \infty$.

Now $g(x) = 1/f(x)$ is also an algebraic function, since

$$p_0(x)(g(x))^n + \cdots + p_{n-1}(x)g(x) + p_n(x) \equiv 0.$$

Hence by exactly the same reasoning, if l is the maximum of $\deg p_j(x) - \deg p_0(x)$, we have

$$\frac{g(x)}{x^{l+1}} \to 0$$

as $x \to \infty$, that is

$$f(x)x^{l+1} \to \infty.$$

Take $n = 1 + \max(k, l)$.

The series expansion

$$2^x = e^{x \ln 2} = 1 + x \ln 2 + \frac{(x \ln 2)^2}{2!} + \cdots + \frac{(x \ln 2)^n}{n!} + \cdots$$

shows that

$$\frac{2^x}{x^n} > \frac{(\ln 2)^n}{n!}$$

for all x and n, and hence the first of these conditions is violated.

The fact that $\sin \pi n = 0$ makes it impossible for us to have $x^n \sin x \to \infty$ as $x \to \infty$.

Question 10.1. If there were an algebraic formula for solving every quintic equation, we could multiply any cubic equation by $(x - r)(x - s)$ and turn it into a quintic, which could then be solved by this formula. That is, the formula would be a multivalued function that assumed all five roots as values when different branches of the radicals were taken. Hence the general cubic $x^3 - ax^2 + bx - c = 0$ could be solved as well using this formula. But the solution of the general cubic requires the extraction of cube roots.

Question 10.2. Abel assumed that the base field containing the coefficients also contained all the roots of unity. If these are adjoined to the real field (in fact, if just the cube roots are joined), then the cubic formula does not go outside the splitting field of the polynomial, that is, all the intermediate computations can be carried out within the field. To see why, not that the cubic formula requires us to adjoin a certain cube root $z = \sqrt[3]{w}$, and the splitting field contains $z + \bar{z}$, which is a root. Since it contains $\bar{a}$, it also contains $\bar{a}z + \bar{a}\bar{z}$. But another root is $az + \bar{a}\bar{z}$, and so the splitting field contains the difference of these two numbers, which is $-z$. Thus, the radical that the cubic formula requires is *inside* the splitting field if that field contains a complex cube root of unity.

Lesson 11

Problem 11.1. We observe first of all that if there are any integers u, v, and w, not all zero, satisfying the relation

$$u^3 + 2v^3 + 4w^3 = 6uvw,$$

then none of them can be zero. The assumption that exactly one of them is zero leads to the conclusion that either $\sqrt[3]{2}$ or $\sqrt[3]{4}$ is a rational number, which we know is not the case; and obviously if two of them are zero, then the third one is also.

If two of the integers u, v, and w satisfying this relation are even, then in fact all three must be even. Obviously 2 does divide u. If it also divides v, then u^3, $2v^3$, and $6uvw$ are all divisible by 8, which implies that $4w^3$ is divisible by 8 and hence that w is also divisible by 2. Likewise, if w is divisible by 2, then u^3, $4w^3$, and $6uvw$ are all divisible by 8, which implies that v is divisible by 2. Thus, if either v or w is even, then in fact all four terms are divisible by 8, and so we can replace u, v, and w by $u' = u/2$, $v' = v/2$, $w' = w/2$ and have the same equation:

$$(u')^3 + 2(v')^3 + 4(w')^3 = 6u'v'w'.$$

Since every nonempty set of nonzero integers contains an integer of minimal absolute value, let w be an integer for which there exist nonzero integers u and v satisfying the given equation and for which $|w|$ is minimal. Then, as just shown, w must be odd. Otherwise, we could cut all three integers in half and get a w with $|w|$ only half as large. Therefore (also by what was just shown), v must also be odd. But this is impossible, since u^3, $4w^3$, and $6uvw$ are all divisible by 4, so that $2v^3$ is divisible by 4.

It follows that there can be no nonzero integer solutions of this equation.

Problem 11.2. Suppose that $r = a/a'$, $s = b/b'$, and $t = c/c'$ (where a, a', b, b', c, and c' are integers) are rational numbers, not all zero, satisfying the relation

$$r^3 + 2s^3 + 4t^3 = 6rst.$$

Then, multiplying through by $(a')^3(b')^3(c')^3$, we find that

$$(ab'c')^3 + 2(a'bc')^3 + 4(a'b'c)^3 = 6(ab'c')(a'bc')(a'b'c).$$

In other words, the integers $u = ab'c'$, $v = a'bc'$, $w = a'b'c$ satisfy the equation in Problem 11.1, which we know is impossible.

Problem 11.3. The splitting field of both equations, by what was shown in the text, is obtained by adjoining $\sqrt{-3}$ (or, equivalently, a complex cube root of unity) and the real number $\sqrt[3]{2}$ to the rational numbers. Hence the splitting field is the same for both. It was shown in the text that the Galois group of automorphisms of this field leaving the rational numbers invariant corresponds to the permutation group S_3.

Problem 11.4. The result of replacing x by $(a + b)$ in the polynomial

$$p(x) = x^6 + 9x^4 - 4x^3 + 27x^2 + 36x + 31$$

and expanding is

$$p(a + b) = 31 + 36a + 27a^2 - 4a^3 + 9a^4 + a^6 + 36b + 54ab - 12a^2b$$
$$+ 36a^3b + 6a^5b + 27b^2 - 12ab^2 + 54a^2b^2 + 15a^4b^2 - 4b^3$$
$$+ 36ab^3 + 20a^3b^3 + 9b^4 + 15a^2b^4 + 6ab^5 + b^6.$$

If you now go through this expression very carefully and replace a^2 by -3, a^3 by $-3a$, a^4 by 9, a^5 by $9a$, a^6 by -27, b^3 by 2, b^4 by $2b$, b^5 by $2b^2$, and b^6 by 4, you will obtain a polynomial in a and b that is formally zero. A computer algebra program will save you some time in this verification.

Problem 11.5. Since we know the roots of this equation in advance, and we know that they are all the complex numbers of the form $a + b$, where $a^2 = 2$ and $b^3 = 2$, it is easy to see that the splitting field contains all three cube roots of 2 and both square roots of 2. For example, since it contains both $\sqrt{2} + \sqrt[3]{2}$ and $-\sqrt{2} + \sqrt[3]{2}$, it must contain $2\sqrt[3]{2}$, and therefore also $\sqrt[3]{2}$, no matter which of the three complex cube roots of 2 this number represents. It follows that the splitting field of this equation contains $(b_1 - b_2)/(b_3)$, where b_1 and b_2 are the two complex cube roots of 2 and b_3 is the real cube root. That is to say, the field contains the difference of the two complex cube roots of unity, which is $\sqrt{-3}$. Hence this field contains both $\sqrt{-3}$ and $\sqrt[3]{2}$, and therefore contains the splitting field $\mathbb{K}$ of the polynomial in the previous problem. But this splitting field also contains $\sqrt{2}$, which is *not* in $\mathbb{K}$. Hence it must be $\mathbb{K}(\sqrt{2})$. The Galois group is therefore a group of 12 elements, containing a copy of Z_2 as a normal subgroup with quotient group S_3. This is not surprising, since the allowable permutations of the roots of this equation are of the form $\sigma\tau$, where σ is a permutation of the two square roots of 2 and τ a permutation of the three cube roots of 2. Let us denote this group by G. It has the same "multiplication table" as the dihedral group D_6 consisting of the symmetries of the regular hexagon, as we shall now show.

Let the vertices of the hexagon be v_0, v_1, v_2, v_3, v_4, and v_5 in cyclic order, as shown in Fig. 18. A symmetry is determined once we specify the locations to which v_0 and v_1 map, and these must be adjacent vertices. Thus there are six possibilities for the image of v_0, and then two for the image of v_1, resulting in a group of order 12. The cyclic permutation $\sigma = (v_0v_1v_2v_3v_4v_5)$ generates the subgroup of rotations, and the permutation $\tau = (v_1v_5)(v_2v_4)$, which leaves v_0 and v_3 fixed, generates a subgroup of order 2. The entire group consists of elements $\sigma^j\tau^k$, $j = 0, 1, 2, 3, 4, 5$, $k = 0, 1$, and $\sigma\tau = (v_1v_0)(v_5v_2)(v_3v_4) = \tau\sigma^5$, as you can easily verify.

The group G can be regarded as the subgroup of S_5 consisting of the permutations of (a, b, c, d, e) leaving the set $\{a, b, c\}$ invariant, that is, moving each element of this set to another element of this set, and hence also leaving $\{d, e\}$ invariant. All we have to do is regard a, b, and c as the cube roots

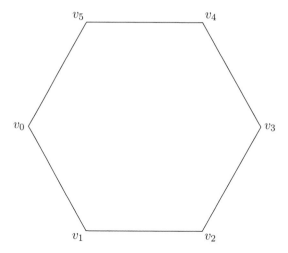

FIGURE 18. A regular hexagon.

of 2 and d and e as the square roots of 2. Any permutation that leaves these two sets invariant will permute the roots of the polynomial and generate an automorphism σ that leaves $\mathbb{Q}$ invariant. A multiplication-preserving correspondence (one of six possible ones) between D_6 and G is established by the mapping $\sigma \mapsto (abc)(de)$ and $\tau \mapsto (ab)$.

Problem 11.6. Most of this problem is straightforward. The three expressions for r^4, r^5, and r^6 are obtained successively from the equation for r^3 my first multiplying the equation for r^n by r, then using the equation for r^3 to eliminate r^3 from the resulting equation. It is then easy to compute that

$$s^3 + s^2 - 2s - 1 = (r^2 - 2)^3 + (r^2 - 2)^2 - 2(r^2 - 2) - 1 = 0$$

and similarly for t, which is $s^2 - 2$, as it happens.

The Cardano solution gives

$$x = -\frac{1}{3} + y = -\frac{1}{3} + \sqrt[3]{\frac{7}{54} + \frac{7\sqrt{3}}{18}i} + \sqrt[3]{\frac{7}{54} - \frac{7\sqrt{3}}{18}i}.$$

It follows that we need to adjoin the cube root of the complex number $\varphi = \frac{7}{54} + \frac{7\sqrt{3}}{18}i$. Since $|\varphi|^2 = \frac{1351}{2916}$, which is a rational number, adjoining φ to the field will result in a larger field that automatically contains $\bar{\varphi} = |\varphi|^2/\varphi$, and hence also the root x. However, the field $\mathbb{Q}(\varphi)$ cannot be obtained from $\mathbb{Q}$ by adjoining just one root of a rational number. Once again, the difficulty comes from not assuming that we start with a field containing the roots of unity. If we started with the field $\mathbb{Q}(\alpha)$, where α is a primitive cube root of unity, we would already have φ in the field, and then the adjunction of its cube root would split the polynomial.

The Viète solution is merely a matter of copying the formula. It expresses the solution in terms of an angle β such that $\cos(\beta) = \frac{1}{2\sqrt{7}}$, but this angle is not a rational multiple of π radians, even though r itself does have a nice expression: $r = 2\cos(2\pi/7)$.

As shown above, since $s = r^2 - 2$, we also have $t = s^2 - 2$. Hence if an automorphism τ that leaves $\mathbb{Q}$ fixed takes r to s, that same automorphism must take s to t. If it leaves r fixed, it must also leave s and t fixed. In both cases, these conclusions follow from the relation $\tau(s) = \tau(r^2 - 2) = \left(\tau(r)\right)^2 - \tau(2) = \left(\tau(r)\right)^2 - 2$.

Problem 11.7. There is only one proper algebraic extension of the real numbers, and that is the complex numbers. (Since the complex numbers are algebraically closed, every polynomial with real coefficients splits completely in the complex numbers.) Its group of automorphisms consists of the identity and complex conjugation; in other words, it is the group Z_2.

Problem 11.8. The solutions of $x^2 + x + 1 = 0$ are cube roots of unity in any field, since $(x - 1)(x^2 + x + 1) = x^3 - 1$. Since both 0 and 1 have square roots, the numbers of the form $a + b\sqrt{c}$ are just the original field of two elements.

Question 11.1. We established above that $\cos 3\theta = 4\cos^3\theta - 3\cos\theta$. Taking $\theta = 20°$, we find that $x = \cos 20°$ satisfies $4x^3 - 3x = \frac{1}{2}$, or $8x^3 - 6x - 1 = 0$. Suppose that we have a set S of points in the plane (complex numbers) that we have been able to locate, starting from 0 and 1, by straightedge-and-compass constructions. Any new point we can locate using ruler and compass may be the intersection of two lines determined by two pairs of points in S, in which case the coordinates of that point are rational functions of the coordinates of the points in S; in other words, they are in the smallest field $\mathbb{F}(S)$ containing all those coordinates.

If not, the new point is the intersection of such a line and a circle with center at a point in S passing through a second point in S. In that case, the coordinates of the new points satisfy a quadratic equation with coefficients in $\mathbb{F}(S)$. Hence their minimal polynomial over $\mathbb{F}(S)$ is of degree 2. In this way, although the rigor leaves something to be desired, we can see that the minimal polynomial over the rational numbers of a Euclidean-constructible number could not be of degree 3. Thus we can see why it is impossible to trisect a 60° angle (construct the cosine of 20°) or double an arbitrary cube (construct $\sqrt[3]{2}$) using ruler and compass. Since π has no minimal polynomial whatsoever over the rational numbers (it is transcendental), we can also see why it is impossible to square the circle using these tools.

Subject Index

abelian group, 153
acceleration, 36, 38, 40, 174
addition, 8
addition formula, 80
Algebra, 27
algebra
  fundamental theorem, 109
  high-school, ix, 52
  linear, 157
  modern, ix, 98
  motivation for, 33–44
  universal, 6
algebra (vector space with
    multiplication), 6, 154, 157–158
algebraic formula, 8, 10, 11, 75,
    118–119, 122, 126, 182
algebraic function, 118–119, 122, 192
algebraic geometry, 10
algebraic number, 8, 9, 11, 125
algebraic operation, 9
algebraic relation, 80
algebraic substitution, 138
algebraically closed field, 9, 10, 62, 109,
    180
algorithm, 21
alternating group, 137
America, 52
analysis, 109, 151
  complex, 10
  input–output, 157
  real, 10
analytic function, 109
analytic geometry, 38
angle trisection, 23, 64, 78, 148
Arabic language, 27, 63
area, 34
arithmetic, 7, 15
arithmetical operations, 8, 109
associative law, 152, 157
asymmetry, 44
automorphism, 128, 129, 136, 138

Banach algebra, 158
binomial theorem, 114
bisection, 45

calculus, 35, 40, 88, 175
  differential, 35
  integral, 35, 36
cancellation law, 152, 153
capital, 33
Cardano formula, 66, 71, 72, 76, 78, 92,
    182, 183
Cardano method, 81, 88
Catalan conjecture, 172
charge, 34, 42
  moving, 43
Chebotarev Density Theorem, 143
chemistry, 18
China, 23, 26, 46, 63, 98, 178
circle, 28, 64, 71, 181
  diameter, 28
coefficient, 10, 11, 18, 22, 23, 57, 58, 77,
    84, 91, 96, 98, 100, 109, 126, 132
  real, 45, 88
Collection of Pappus, 6
commutative group, 153
complex analysis, 10
complex conjugate, 110, 138
complex number, 8, 11, 19, 45, 62, 71,
    76, 81, 92, 109, 136, 179
  cube root, 63, 64, 67–68, 79
  imaginary part, 59, 77, 79
  real part, 59
  square root, 59
computer, 58
computer algebra, 91, 102
conic section, 63
conjugate, 137
  complex, 138
conjugate radical, 81
conjugation, 127
continued fraction, 163
continuity, 110

Copernican theory, 42
coset, 129
    left, 130
    of an ideal, 155
    right, 130
cosine, 79
coulomb (unit of charge), 43
Coulomb's law, 42
counter, 48
counting board, 47, 48
Crime and Punishment, 159
cross product, 13
    anticommutativity, 13
crossword puzzle, 34
crystal, 154
cube root, 9, 63, 76, 87, 130, 139, 177,
        179, 182
    of a complex number, 64, 67–68, 79,
        183
    of a real number, 63
    of unity, 68, 76, 84
cubic discriminant, 70, 182
cubic equation, 17, 28, 51, 58, 63–72,
        75–88, 104, 112, 126, 180–182
    graphical solution, 63
    irreducible case, 69–70, 76–79, 122
    Tschirnhaus solution, 80–82, 88
cubic formula, 67–69, 112, 116–117,
        131, 137
    in terms of roots, 83–85
cubic polynomial, 82, 87
current, 34
cycle, 102
    2-, 103, 138, 190
    3-, 102, 103, 190
    4-, 103, 190
    5-, 103, 138, 190
    order, 190
cyclic group, 137, 154
cyclotomic equation, 87, 127, 187

decimal expansion, 10, 23, 57
degrees of freedom, 19, 23
density, 34
derivative, 88
Descartes' rule of signs, 137, 163–165,
        172, 185
determinant, 48, 152
determinate system, 172
dielectric permittivity, 42
differential calculus, 35
differential equation, xii, 35, 37, 152,
        154

Diophantine equation, 20, 23, 24, 147
dirhem, 27
discriminant, 58, 61, 69, 101
    cubic, 70, 78, 79, 81, 182
    quadratic, 58, 77, 110
        negative, 58
distance, 34
distributive law, 154, 157
division, 8, 25, 139
    synthetic, 52
division ring, 156
dot product, 13
double root, 70, 88, 188
doubling the cube, 23, 63, 64
ducat, 33

e, 11
Earth, 41, 42
    radius, 41
$E = mc^2$, 35
Egypt, 25, 46
electric field, 43, 176
electromagnetic theory, 43
electromagnetic wave, 43, 175
Elements of Euclid, 163
ellipse, 118
elliptic function, xii, 35, 80
elliptic integral, 80
energy, 34
equation, 11
    cubic, 17, 28, 51, 58, 61, 63–72,
        75–88, 104, 112, 126, 180–182
        Cardano solution, 80
        irreducible case, 69–70, 76–79, 122
        resolvent, 77
        Tschirnhaus solution, 80–82, 88
    cyclotomic, 87, 127, 187
    determinate system, 19, 23, 24
    differential, xii, 35, 37, 152, 154
    Diophantine, 20, 23, 24, 147, 151
    formulaic solution, 151
    heat, 35
    ideal gas, 34
    indeterminate system, 19, 23, 24
    linear, 28, 76
    linear system, 22, 48, 61, 84
    overdetermined system, 20, 23, 24
    Pell's, 20
    polynomial, xi, 17, 23
    "Pythagorean", 20
    quadratic, 17, 19, 21, 24, 26, 28, 29,
        57–62, 76, 77, 82, 83, 86, 110,
        167, 185

quartic, 65, 76–78, 84–86, 93, 97, 101, 142, 184, 189
quintic, 75, 91, 92, 95, 96, 109, 125, 137
  transcendental solution, 80
  unsolvability, 112–122
resolvent, 78, 88, 92
Schrödinger, 35
sextic, 146
solution, 8, 17, 45, 135
  Chinese method, 49–52, 63, 78
  formulaic, 21–23, 30, 57–71, 125, 151
  numerical, 21, 23, 30, 63
  real, 62, 77, 179
  tabular, 57–60, 63–64
system
  determinate, 19, 23, 24
  indeterminate, 19, 23, 24
  linear, 22, 48, 61, 84
  overdetermined, 20, 23, 24
wave, 35
equator, 41
Euclidean algorithm, 161–163
Euclidean space, 151, 156
Europe, 23, 34, 64
exponential, 35
extraneous root, 82

false position, 25
Fermat's Last Theorem, 119
field, ix, 6, 8, 75, 132, 156, 169–170
  algebraically closed, 9, 10, 62, 180
  electric, 43, 176
  finite, 8, 29, 148, 180, 196
  gravitational, 38
  magnetic, 43, 176
  of five elements, 13, 24, 52, 171, 173, 177
  of four elements, 12
  of three elements, 12, 24, 29, 133, 167, 171, 172
  of two elements, 12, 30, 62, 148, 173, 196
  splitting, 120, 125, 147
field extension, 11
fifth root of unity, 87, 88, 188
finite field, 8, 29, 148, 180, 196
5-cycle, 103, 138, 190
five-element field, 171, 173, 177
force, 42, 43
formula

algebraic, 8, 10, 11, 60, 75, 118–119, 122, 126, 182
Cardano, 66, 71, 72, 76, 78, 182, 183
cubic, 67–69, 112, 116–117, 131, 137
  in terms of roots, 83–85
double-valued, 60
quadratic, 29, 58, 60–61, 112–113, 172, 178
formulaic solution, 125
4-cycle, 103, 190
Fourier transform, 158
fourth root, 9
  of unity, 182
fraction, 8, 13
function
  algebraic, 58, 118–119, 122, 192
  analytic, 109
  elliptic, xii, 35, 80
  Jacobi amplitude, 35, 80
  nonsymmetric, 61
  rational, 58, 60, 97, 99, 132, 182
  single-valued, 60
  symmetric, 58, 61, 84, 91, 100, 121
    elementary, 58, 100
  three-valued, 101, 102
  transcendental, 122
  two-valued, 100–101
fundamental theorem of algebra, 109
fundamental theorem of arithmetic, 156

g (acceleration of gravity), 36, 39, 41
Galois group, 128, 131, 136
Galois theory, ix, xii, 10, 23, 91, 98, 125–149, 152, 193–196
Gaussian domain, 156
Gaussian integer, 151, 154, 156
geese, 33
geometry, 6, 10, 11, 18–19, 23, 26, 28, 109, 151
  analytic, 38
$GL(2, \mathbb{R})$ (general linear group), 152
grain, 47
gravitational, 35, 175
gravitational field, 38
greatest common divisor, 161–163
Greeks, 23, 46
group, ix, xii, 6, 75, 98, 128, 138, 151–154
  abelian, 153
  alternating, 129, 137
  commutative, 153
  cyclic, 137, 138, 154
  dihedral, 194

Galois, ix, 128, 131, 136
  Klein, 142
  quotient, 129, 136
  solvable, 137
  symmetric, 138, 152
group theory, xii, 18

Han Dynasty, 26
heat equation, 35
Hindu–Arabic numerals, 35
Hindus, 33, 63, 98
Horner's method, 52
hyperbola, 18, 63, 71, 110, 172, 181
  degenerate, 111

ideal, 158
  in a ring, 155
  maximal, 158
ideal gas, 34
identity
  in a group, 152
  in a ring, 156
  polarization, 22
imaginary number, 9, 58, 70–71
imaginary part of a complex number,
    59, 77, 79
impossibility proof, 23
independence, 20
indeterminate system, 172
index of a subgroup, 130
India, 7, 34
induction, magnetic, 42
inertia, 39
infinitesimal methods, 35, 38, 40
inheritance problem, 28
inner product, 13
input–output analysis, 157
integer, 8, 11, 24, 125, 154
  even, 156
  Gaussian, 154, 156
  negative, 8, 9
integral calculus, 35, 36
integral domain, 156, 163
  Euclidean, 163
inverse of a group element, 152
inversion, 99
Iraq, 26
irrational number, 9, 10
irreducible element, 156
isomorphism, 136
Italy, 23, 64, 65

Jacobi amplitude, 35, 37, 80
Jains, 7

Japan, 23, 28, 98

Kepler's third law, 34, 42
Klein four-group, 142

latus rectum, 118
law
  Coulomb's, 42
  inverse-square, 38, 42
  Kepler's third, 34, 42, 175
  of inertia, 39
  of universal gravitation, 35
  Ohm's, 35
  Stefan–Boltzmann, 35
left coset, 130
Les misérables, 133
light, 43, 175
  speed of, 43
like charges, 42
limit, 110
line, 18, 64
linear algebra, 47, 157
linear equation, 28, 76
linear system, 101
Louvre, 26

magnetic field, 43, 176
magnetic induction, 42
magnetic permeability, 43
manifold, 151
Maple, 91, 143, 144
mass, 34
Mathematica, 91, 144
Mathematical Association of America, x
Mathematical Capsules, x
Matlab, 91
matrix, 47, 48, 152, 157
  upper-triangular, 154
  Vandermonde, 49, 61, 84, 96, 101, 120
Maxwell's laws, 175
mechanics
  Newtonian, 43
medieval Muslims, 23, 64, 65
Merton College, 39
Merton rule, 39, 40
Mesopotamia, 22, 26, 64
meter, 43
method
  Cardano, 88
  Tschirnhaus, 87, 184–187
  Viète's, 79, 86, 184
method of false position, 25
Möbius transformation, 86
modern algebra, 98

module, 6, 157
  unitary, 157
Moon, 38, 41
  orbital radius, 42
motion
  at constant velocity, 38
  uniformly accelerated, 38
multiplication, 8, 25

negative number, 65
newton (unit of force), 43
Newton's law of universal gravitation,
  35
Newton's laws of motion, 36
Newton–Raphson approximation, 59
Newtonian mechanics, 43
*Nine Chapters on the Mathematical
  Art*, 26, 47
norm, 13
normal subgroup, 130, 153
North Pole, 41
number
  algebraic, 8, 9, 11, 125
  Chinese notation, 46–47
  complex, 8, 11, 19, 45, 62, 71, 76, 81,
    92, 109, 136, 179
    cube root, 63, 64, 67–68, 79, 182,
      183
    imaginary part, 59, 79
    real part, 59
    square root, 59, 182
  imaginary, 9, 58, 70–71
  irrational, 9, 10
  negative, 65
  prime, 121
  rational, 8, 11, 21, 87, 110, 125, 136
  real, 8, 11, 19, 21, 45, 58, 110, 138,
    152, 180
    cube root, 63
  transcendental, 10, 11
number theory, 151
numerals, 35
  Hindu–Arabic, 35
  Roman, 35
numerical approximation, 21, 46
numerical solution, 21, 63, 176–178

observer, 43
Ohm's law, 35
operation
  arithmetical, 109
  rational, 11, 13, 125, 136
order of a group element, 138
Oxford, 39, 158

Pacific Journal of Mathematics, 137
parabola, 63
parameter, 63, 81
Paris, 26
parity of a permutation, 99
Pell's equation, 20
pendulum, 35–37
period
  sidereal, 41
permanence of functional relations, 116
permeability, magnetic, 43
permittivity, dielectric, 42
permutation, 48, 91, 97–101, 138, 152,
    154, 189–191
  cyclic, 102, 129
  even, 129, 130
  identity, 190
  inverse, 130
  odd, 130
  order, 103, 190
  parity, 99
perpendicularity, 14
physics, 18, 34–44
π, 10, 11
planet, 34
  distance from the Sun, 42
  period, 42
polarization identity, 22
polynomial, ix, x, 17, 110, 157
  cubic, 82, 87
  splitting field, 125
polynomial equation, xi, 23
polynomial ring, 154, 163
Prague Scientific Society, 121
pressure, 34
prime element of an integral domain,
    156
prime number, 121
primitive root of unity, 87, 182
profit, 33
puzzle
  crossword, 34
  sliding-frame, 103
"Pythagorean" equation, 20

quadratic equation, 12, 17, 19, 21, 24,
    26, 28, 29, 57–62, 76, 77, 82, 83, 86,
    110, 167, 185
quadratic formula, 29, 58, 60–61,
    112–113, 172, 178
quartic equation, 65, 76–78, 84–86, 93,
    97, 101, 142, 184
quaternion, 13–14, 62, 156, 169, 180

quintic equation, 75, 91, 92, 95, 96, 109,
        125, 137
    transcendental solution, 80
    unsolvability, 112–122
quotient, 25
quotient group, 129, 136

radical, 81
    conjugate, 81
rational function, 60, 97, 99, 132, 182
rational number, 8, 11, 21, 87, 110, 125,
        136
rational operation, 8, 11, 13, 125, 136
real analysis, 10
real number, 8, 11, 19, 21, 45, 58, 110,
        138, 152, 180
    cube root, 63
real part of a complex number, 59
rectangle, 26, 38
relation
    algebraic, 80
    transcendental, 80
relativity, 176
resolvent, 91, 92
    for the quartic, 94
resolvent equation, 88
Rhind Mathematical Papyrus, 25
Riemann surface, 114
right coset, 130
ring, ix, 6, 154–156
    associative, 154–155
    Lie, 155–156
    polynomial, 154, 163
Rolle's theorem, 164
Roman numerals, 35
root, 22, 50, 57, 58, 84, 91, 97, 100, 109
    nth, 76
    approximate, 177
    cube, 139, 177, 179
        of a complex number, 182
    double, 70, 88, 188
    extraneous, 82
    finding, 45
    of unity, 71, 87, 96, 120, 126, 173, 182
        fifth, 188
        fourth, 182
        primitive, 68, 87, 182
    rational, 161
    single, 70
    square, 172, 176, 179, 196
        of a complex number, 182
    triple, 70

root extraction, 8, 9, 18, 61, 109, 125,
        137
root of unity, 132–133
    primitive, 132

scalar, 156
Schrödinger equation, 35
Schwarz inequality, 13
set theory, 151
sextic equation, 146
sidereal period, 41
single root, 70
sliding-frame puzzle, 103
solution, 10
    algebraic, 135, 138
    Chinese method, 59
    formulaic, 21–23, 30
    numerical, 23, 30, 176–178
    real, 79
    transcendental form, 80
solvable group, 137
space
    Euclidean, 151
special relativity, 44
speed of light, 175
sphere, 41
    area, 41
splitting field, 120, 125, 147
square root, 9, 22, 29, 33, 58, 61, 81, 87,
        172, 176, 179, 182, 196
    of a complex number, 59
    table, 57
squaring the circle, 23
Stefan–Boltzmann law, 35
straightedge-and-compass
        constructions, 23, 148
string theory, 157
subfield, 11
subgroup, 129, 134, 151, 153
    index, 130
    nonnormal, 151
    normal, 130, 134, 151, 153
substitution, algebraic, 138
subtraction, 8, 167
Sun, 34
superstring theory, 157
symmetric function, 58, 84, 91, 100, 121
    elementary, 58, 100
symmetric group, 138, 152
symmetrizing, 81
symmetry, 18, 40, 58
    breaking, 18, 22, 70
synthetic division, 52

system
   determinate, 172
   indeterminate, 172

temperature, 34
The Harmonies of the World, 34
theory
   Copernican, 42
   electromagnetic, 43
3-cycle, 102, 103, 137, 190
three-element field, 133, 167, 171, 172
three-valued function, 101, 102
time, 34
topology, 110, 151
transcendental function, 122
transcendental number, 10, 11
transcendental relation, 80
transformation
   fractional-linear, 86
   Möbius, 86
transposition, 99, 100, 103
trigonometric series, 151
trigonometry, xii, 35, 40, 71, 76, 78, 88,
   187
triple root, 70
trisection, 148
Tschirnhaus method, 82, 87, 131,
   184–187
2-cycle (transposition), 103, 138, 190
two-element field, 173

two-valued function, 100–101

unique factorization domain, 156
unit fractions, 25
unit in a ring, 156, 163
unitary law, 157
universal algebra, 6
unknown, 11, 17

Vandermonde matrix, 49, 61, 96, 101,
   120
variable, x, 34
   complex, 109
   real, 72
vector, xii, 13, 156, 167–169
   absolute value, 13
   length, 13
   norm, 13
vector space, xii, 6, 154, 156–157
   four-dimensional, 13
Viète's method, 79, 86, 184
volume, 34

wave equation, 35
wave, electromagnetic, 43, 175
weber (unit of magnetic induction), 43
wheat, 47

zero, 8, 47
zero divisor, 155

Name Index

Abel, Niels Henrik (1802–1829), 92, 109, 112, 113, 119, 120, 125, 131, 137
Apollonius of Perga (third century BCE), 64, 118
Aristotle (fourth century BCE), 176

Bach, Johann Sebastian (1685–1750), ix
van Beethoven, Ludwig (1770–1827), ix
Bell, Eric Temple (1883–1960), 154
Bhaskara II (1114–1185), 33
Bochner, Salomon (1899–1982), 158
Boltzmann, Ludwig (1844–1906), 35, 176
Bombelli, Rafael (1526–1572), 66, 70, 76, 78, 92, 104, 182
Brahmagupta (seventh-century), 20

Cantor, Georg (1845–1918), 151
Cardano, Girolamo (1501–1576), xi, 33, 66, 71, 72, 76, 78, 81, 85, 88, 92, 148, 182, 183, 185, 195
Catalan, Eugène (1814–1894), 172
Cauchy, Augustin-Louis (1789–1856), 101, 109, 113, 121, 151
Cayley, Arthur (1821–1895), x
Chebotarev, Nikolai Grigorevich (1894–1947), 143
Ch'in Shih Huang Ti (259–209 BCE), Chinese Emperor, 3
Copernicus, Nicolaus (1473–1543), 42
Coulomb, Charles (1736–1806), 42

Descartes, René (1596–1650), 34, 39, 137, 163, 164, 172, 185
Diophantus of Alexandria (third century), 20, 23, 24, 147, 151
Dostoevsky, Fyodor (1821–1881), 159
Dummit, David, 144

Einstein, Albert (1879–1957), 35
Euclid of Alexandria (fl. ca. 300 BCE), 149, 163
Euler, Leonhard (1707–1783), x, 34, 92

Feit, Walter (1930–2004), 137
de Fermat, Pierre (1601–1665), 119
Ferrari, Ludovico (1522–1565), 66, 76
del Ferro, Scipione (1465–1525), 65
Fraenkel, Adolf (1891–1965), 154

Galilei, Galileo (1564–1642), 39
Galois, Évariste (1811–1832), ix, xii, 10, 23, 91, 98, 118, 125–149, 151, 193–196
Gauss, Carl Friedrich (1777–1855), 109, 151, 156, 164
Gel'fand, Izrail' Moiseevich (b. 1913), 158
Gibbs, Josiah Willard (1839–1903), 13, 14
Grassmann, Hermann Günter (1809–1877), x
Grattan-Guinness, Ivor, x

Hamilton, William Rowan (1805–1865), 13, 14, 122
Hermite, Charles (1822–1900), 122
Horner, William George (1787–1837), 52
Hugo, Victor (1802–1885), 133

Jacobi, Carl Gustav (1804–1851), 35, 37, 80, 155
Jardine, Richard, x

Kepler, Johannes (1571–1630), 34, 42, 175
al-Khayyam, Umar (1048–1131), xi, 63, 64
al-Khwarizmi, Muhammed ibn-Musa (ca. 790–840), xi, 27
Klein, Felix (1849–1925), 142, 152, 158
Kronecker, Leopold (1823–1891), 120, 122
Kummer, Ernst Eduard (1810–1893), 151

Lagrange, Joseph-Louis (1736–1813), 82–83, 85, 92, 94, 96, 97, 104, 120
Lazard, Daniel, 144
von Leibniz, Gottfried (1646–1716), 76, 98
Leontief, Wassily (1906–1999), 157
Lie, Sophus (1842–1899), 152, 155

Maxwell, James Clerk (1832–1879), 43, 175
Mermin, N. David (b. 1935), 176
Mihăilescu, Preda (b. 1955), 172
Möbius, August Ferdinand (1790–1868), 86

Newton, Isaac (1642–1727), 35, 36, 42, 43, 201
Nový, Luboš (b. 1929), x

Ohm, Georg Simon (1789–1854), 35

Pappus of Alexandria (ca. 290–ca. 350), 6, 143
Pell, John (1611–1685), 20
Pesic, Peter, x
Poincaré, Henri (1854–1912), 176
Puiseux, Victor (1820–1883), 113

Raphson, Joseph (1648–1715), 201
Riemann, Bernhard (1826–1866), 114, 151
Rolle, Michel (1652–1719), 164
Rosen, Michael, 131
Ruffini, Paolo (1765–1822), 96, 109, 121

Sawaguchi, Kazuyuki (dates uncertain), 28, 173
Schrödinger, Erwin (1887–1961), 35
Schwarz, Hermann Amandus (1843–1921), 13
Seki, Kowa (1642–1708), 52
Seki, Takakazu (1642–1708), 52
Shell-Gellasch, Amy, x
Stefan, Josef (1835–1893), 35
Sylow, Peter (1832–1918), 138
Sylvester, James Joseph (1814–1897), x

Tartaglia, Niccolò (1500–1557), 66
Tee, Garry J., xii
Thompson, John (b. 1932), 137
Tignol, Jean-Pierre, x
von Tschirnhaus, Ehrenfried Walther (1652–1708), 76, 80–82, 87, 88, 92, 96, 131, 184–187
al-Tusi, Sharaf al-Din al-Muzaffar (ca. 1135–1213), 63, 65

Vandermonde, Alexandre (1735–1796), x, 49, 84, 96, 101, 120
Viète, François (1540–1603), 34, 71, 76, 78, 79, 86, 88, 119, 122, 148, 184, 196

Waring, Edward (1736–1798), x
Weber, Wilhelm (1804–1891), 43
Wiener, Norbert (1894–1964), 158